Lecture Notes in Mathematics 849

For further volumes:
http://www.springer.com/series/304

Péter Major

Multiple Wiener-Itô Integrals

With Applications to Limit Theorems

Second Edition

Springer

Péter Major
Hungarian Academy of Sciences
Alfréd Rényi Mathematical Institute
Budapest, Hungary

ISBN 978-3-319-02641-1 ISBN 978-3-319-02642-8 (eBook)
DOI 10.1007/978-3-319-02642-8
Springer Cham Heidelberg New York Dordrecht London

Lecture Notes in Mathematics ISSN print edition: 0075-8434
ISSN electronic edition: 1617-9692

Library of Congress Control Number: 2013953865

Mathematics Subject Classification (2010): Primary: 60G18, 60H05, 60F99,
secondary: 60G10, 60G15, 60G60

Printed on acid-free paper

Springer is part of Springer Science+Business Media (www.springer.com)

Preface to the Second Edition

This text is a slightly modified version of my Lecture Note *Multiple Wiener–Itô integrals with applications to limit theorems* published in the *Lecture Notes in Mathematics* series (number 849) of the Springer Verlag in 1981. I decided to write a revised version of this Lecture Note after a special course I held about its subject in the first semester of the academic year 2011–2012 at the University of Szeged. Preparing for this course I observed how difficult the reading of formulas in this Lecture Note was. These difficulties arose because this Lecture Note was written at the time when the TEX program still did not exist, and the highest technical level of typing was writing on an IBM machine that enabled one to type beside the usual text also mathematical formulas. But the texts written in such a way are very hard to read. To make my text more readable, I decided to retype it by means of the TEX program. But it turned out that a real improvement of the text demands much more than producing nice, readable formulas. To make a really better version of this work, I also had to explain better the results and definitions together with the ideas and motivation behind them. Besides, I had to make not only more readable formulas but also more readable explanations. The reader must see at each point of the discussion what is just going on and why. In the new version of this work, I tried to satisfy these demands. Naturally, I also corrected the errors I found. At some points I had to insert a rather long explanation in the proof, because I met such a statement which seemed to be trivial at the first sight, but its justification demanded a detailed discussion. I hope that these insertions did not make the work less transparent.

There appeared many new results about the subject of this Lecture Note since its first appearance. The question arose naturally whether I should insert them to the new edition of this work. Finally I decided to make no essential changes in the text, to restrict myself to the correction of the errors I found and to give a more detailed explanation of the proofs where I felt that it is useful. In making such a decision I

The preparation of this revised version was supported by the Hungarian National Foundation for Science (OTKA) under grant number 91928.

was influenced by a Russian proverb which says: “Лучше враг хорошего”. I tried to follow the advice of this proverb. (I do not know of an English counterpart of this proverb, but it has a French version: “Le mieux est l’ennemi du bien”.)

I made one exception. I decided to explain those basic notions and results in the theory of generalized functions which were applied in the older version of this work in an implicit way. In particular, I tried to explain with their help how one gets those results about the spectral representation of the covariance function of stationary random fields that I have presented under the names *Bochner’s theorem* and *Bochner–Schwartz theorem*. This extension of the text is contained in the attachments to Chaps. 1 and 3. In the first version I only referred to a work where these notions and results can be found. But now I found such an approach not satisfactory, because these notions and results play an important role in some arguments of this work. Hence I felt that to make a self-contained presentation of the subject, I have to explain them in more detail.

The first edition of this Lecture Note appeared long time ago, but the main question discussed in it, the description of the limit behaviour of appropriately normalized partial sums of strongly dependent random variables remained an open problem. Also the method applied in this work remained an important tool in the study of such problems. Hence a self-contained explanation of the theory which provides a good foundation for this method is useful. By my hopes this Lecture Note contains such an explanation, and therefore it did not become out of date. This was the main argument for myself to write a new version of this work where I tried to present a better and more accessible discussion.

I would like to write some words about the last chapter of this work, where some results are discussed that seemed to be important at the time of writing the first version. I would mention two of them which later turned out to be really important. The first one is the Nelson–Gross inequality which later played an important role in the theory of the so-called hypercontractive and logarithmic Soboliev inequalities. The second one is a method for construction of non-trivial self-similar fields worked out in a paper of Kesten and Spitzer. Several important limit theorems are based on the ideas of this paper. It is worth mentioning that it was Roland L’vovich Dobrushin who called my attention to these results, and he emphasized their importance. So I would like to finish this preface with a personal remark about him.

This work is the result of some joint research with Roland L’vovich Dobrushin. Although the book was written by me alone, Dobrushin’s influence is very strong in it. I have learned very much from him. It is rather difficult to explain what one could learn from him, because it was much more than just some results or mathematical arguments. There was something beyond it, some world view which is hard to explain. If I could give back something from what I had learned from him in this Lecture Note, then this would justify the work on it by itself.

Budapest, Hungary
15 August 2013

Péter Major

Preface to the Second Edition

This text is a slightly modified version of my Lecture Note *Multiple Wiener–Itô integrals with applications to limit theorems* published in the *Lecture Notes in Mathematics* series (number 849) of the Springer Verlag in 1981. I decided to write a revised version of this Lecture Note after a special course I held about its subject in the first semester of the academic year 2011–2012 at the University of Szeged. Preparing for this course I observed how difficult the reading of formulas in this Lecture Note was. These difficulties arose because this Lecture Note was written at the time when the TEX program still did not exist, and the highest technical level of typing was writing on an IBM machine that enabled one to type beside the usual text also mathematical formulas. But the texts written in such a way are very hard to read. To make my text more readable, I decided to retype it by means of the TEX program. But it turned out that a real improvement of the text demands much more than producing nice, readable formulas. To make a really better version of this work, I also had to explain better the results and definitions together with the ideas and motivation behind them. Besides, I had to make not only more readable formulas but also more readable explanations. The reader must see at each point of the discussion what is just going on and why. In the new version of this work, I tried to satisfy these demands. Naturally, I also corrected the errors I found. At some points I had to insert a rather long explanation in the proof, because I met such a statement which seemed to be trivial at the first sight, but its justification demanded a detailed discussion. I hope that these insertions did not make the work less transparent.

There appeared many new results about the subject of this Lecture Note since its first appearance. The question arose naturally whether I should insert them to the new edition of this work. Finally I decided to make no essential changes in the text, to restrict myself to the correction of the errors I found and to give a more detailed explanation of the proofs where I felt that it is useful. In making such a decision I

The preparation of this revised version was supported by the Hungarian National Foundation for Science (OTKA) under grant number 91928.

was influenced by a Russian proverb which says: “Лучше враг хорошего”. I tried to follow the advice of this proverb. (I do not know of an English counterpart of this proverb, but it has a French version: “Le mieux est l’ennemi du bien”.)

I made one exception. I decided to explain those basic notions and results in the theory of generalized functions which were applied in the older version of this work in an implicit way. In particular, I tried to explain with their help how one gets those results about the spectral representation of the covariance function of stationary random fields that I have presented under the names *Bochner’s theorem* and *Bochner–Schwartz theorem*. This extension of the text is contained in the attachments to Chaps. 1 and 3. In the first version I only referred to a work where these notions and results can be found. But now I found such an approach not satisfactory, because these notions and results play an important role in some arguments of this work. Hence I felt that to make a self-contained presentation of the subject, I have to explain them in more detail.

The first edition of this Lecture Note appeared long time ago, but the main question discussed in it, the description of the limit behaviour of appropriately normalized partial sums of strongly dependent random variables remained an open problem. Also the method applied in this work remained an important tool in the study of such problems. Hence a self-contained explanation of the theory which provides a good foundation for this method is useful. By my hopes this Lecture Note contains such an explanation, and therefore it did not become out of date. This was the main argument for myself to write a new version of this work where I tried to present a better and more accessible discussion.

I would like to write some words about the last chapter of this work, where some results are discussed that seemed to be important at the time of writing the first version. I would mention two of them which later turned out to be really important. The first one is the Nelson–Gross inequality which later played an important role in the theory of the so-called hypercontractive and logarithmic Soboliev inequalities. The second one is a method for construction of non-trivial self-similar fields worked out in a paper of Kesten and Spitzer. Several important limit theorems are based on the ideas of this paper. It is worth mentioning that it was Roland L’vovich Dobrushin who called my attention to these results, and he emphasized their importance. So I would like to finish this preface with a personal remark about him.

This work is the result of some joint research with Roland L’vovich Dobrushin. Although the book was written by me alone, Dobrushin’s influence is very strong in it. I have learned very much from him. It is rather difficult to explain what one could learn from him, because it was much more than just some results or mathematical arguments. There was something beyond it, some world view which is hard to explain. If I could give back something from what I had learned from him in this Lecture Note, then this would justify the work on it by itself.

Budapest, Hungary
15 August 2013

Péter Major

Preface to the Second Edition

This text is a slightly modified version of my Lecture Note *Multiple Wiener–Itô integrals with applications to limit theorems* published in the *Lecture Notes in Mathematics* series (number 849) of the Springer Verlag in 1981. I decided to write a revised version of this Lecture Note after a special course I held about its subject in the first semester of the academic year 2011–2012 at the University of Szeged. Preparing for this course I observed how difficult the reading of formulas in this Lecture Note was. These difficulties arose because this Lecture Note was written at the time when the TeX program still did not exist, and the highest technical level of typing was writing on an IBM machine that enabled one to type beside the usual text also mathematical formulas. But the texts written in such a way are very hard to read. To make my text more readable, I decided to retype it by means of the TeX program. But it turned out that a real improvement of the text demands much more than producing nice, readable formulas. To make a really better version of this work, I also had to explain better the results and definitions together with the ideas and motivation behind them. Besides, I had to make not only more readable formulas but also more readable explanations. The reader must see at each point of the discussion what is just going on and why. In the new version of this work, I tried to satisfy these demands. Naturally, I also corrected the errors I found. At some points I had to insert a rather long explanation in the proof, because I met such a statement which seemed to be trivial at the first sight, but its justification demanded a detailed discussion. I hope that these insertions did not make the work less transparent.

There appeared many new results about the subject of this Lecture Note since its first appearance. The question arose naturally whether I should insert them to the new edition of this work. Finally I decided to make no essential changes in the text, to restrict myself to the correction of the errors I found and to give a more detailed explanation of the proofs where I felt that it is useful. In making such a decision I

The preparation of this revised version was supported by the Hungarian National Foundation for Science (OTKA) under grant number 91928.

was influenced by a Russian proverb which says: “Лучше враг хорошего”. I tried to follow the advice of this proverb. (I do not know of an English counterpart of this proverb, but it has a French version: “Le mieux est l’ennemi du bien”.)

I made one exception. I decided to explain those basic notions and results in the theory of generalized functions which were applied in the older version of this work in an implicit way. In particular, I tried to explain with their help how one gets those results about the spectral representation of the covariance function of stationary random fields that I have presented under the names *Bochner’s theorem* and *Bochner–Schwartz theorem.* This extension of the text is contained in the attachments to Chaps. 1 and 3. In the first version I only referred to a work where these notions and results can be found. But now I found such an approach not satisfactory, because these notions and results play an important role in some arguments of this work. Hence I felt that to make a self-contained presentation of the subject, I have to explain them in more detail.

The first edition of this Lecture Note appeared long time ago, but the main question discussed in it, the description of the limit behaviour of appropriately normalized partial sums of strongly dependent random variables remained an open problem. Also the method applied in this work remained an important tool in the study of such problems. Hence a self-contained explanation of the theory which provides a good foundation for this method is useful. By my hopes this Lecture Note contains such an explanation, and therefore it did not become out of date. This was the main argument for myself to write a new version of this work where I tried to present a better and more accessible discussion.

I would like to write some words about the last chapter of this work, where some results are discussed that seemed to be important at the time of writing the first version. I would mention two of them which later turned out to be really important. The first one is the Nelson–Gross inequality which later played an important role in the theory of the so-called hypercontractive and logarithmic Soboliev inequalities. The second one is a method for construction of non-trivial self-similar fields worked out in a paper of Kesten and Spitzer. Several important limit theorems are based on the ideas of this paper. It is worth mentioning that it was Roland L’vovich Dobrushin who called my attention to these results, and he emphasized their importance. So I would like to finish this preface with a personal remark about him.

This work is the result of some joint research with Roland L’vovich Dobrushin. Although the book was written by me alone, Dobrushin’s influence is very strong in it. I have learned very much from him. It is rather difficult to explain what one could learn from him, because it was much more than just some results or mathematical arguments. There was something beyond it, some world view which is hard to explain. If I could give back something from what I had learned from him in this Lecture Note, then this would justify the work on it by itself.

Budapest, Hungary
15 August 2013

Péter Major

Preface to the First Edition

One of the most important problems in probability theory is the investigation of the limit distribution of partial sums of appropriately normalized random variables. The case where the random variables are independent is fairly well understood. Many results are known also in the case where independence is replaced by an appropriate mixing condition or some other "almost independence" property. Much less is known about the limit behaviour of partial sums of really dependent random variables. On the other hand, this case is becoming more and more important, not only in probability theory but also in some applications in statistical physics.

The problem about the asymptotic behaviour of partial sums of dependent random variables leads to the investigation of some very complicated transformations of probability measures. The classical methods of probability theory do not seem to work for this problem. On the other hand, although we are still very far from a satisfactory solution of this problem, we can already present some non-trivial results.

The so-called multiple Wiener–Itô integrals have proved to be a very useful tool in the investigation of this problem. The proofs of almost all rigorous results in this field are closely related to this technique. The notion of multiple Wiener–Itô integrals was worked out for the investigation of non-linear functionals over Gaussian fields. It is closely related to the so-called Wick polynomials which can be considered as the multi-dimensional generalization of Hermite polynomials. The notion of Wick polynomials and multiple Wiener–Itô integrals were worked out at the same time and independently of each other. Actually, we discuss a modified version of the multiple Wiener–Itô integrals in greatest detail. The technical changes needed in the definition of these modified integrals are not essential. On the other hand, these modified integrals are more appropriate for certain investigations, since they enable us to describe the action of shift transformations and to apply some sort of random Fourier analysis. There is also some connection between multiple Wiener–Itô integrals and the classical stochastic Itô integrals. The main difference between them is that in the first case deterministic functions are integrated, and in the second case so-called non-anticipating functionals. The consequence of this difference is that no technical difficulty arises when we want to define multiple

Wiener–Itô integrals in the multi-dimensional time case. On the other hand, a large class of non-linear functionals over Gaussian fields can be represented by means of multiple Wiener–Itô integrals.

In this work we are interested in limit problems for sums of dependent random variables. It is useful to consider this problem together with its continuous time version. The natural formulation of the continuous time version of this problem can be given by means of generalized random fields. Consequently we also have to discuss some questions about them.

I have not tried to formulate all the results in the most general form. My main goal was to work out the most important techniques needed in the investigation of such problems. This is the reason why the greatest part of this work deals with multiple Wiener–Itô integrals. I have tried to give a self-contained exposition of this subject and also to explain the motivation behind the results.

I had the opportunity to participate in the Dobrushin–Sinai seminar in Moscow. What I learned there was very useful also for the preparation of this Lecture Note. Therefore I would like to thank the members of this seminar for what I could learn from them, especially P.M. Bleher, R.L. Dobrushin and Ya.G. Sinai.

Contents

Acronyms

$\mathscr{D}$	The space of infinitely differentiable functions with compact support
$\mathscr{D}'$	The space of generalized functions on the space of test function $\mathscr{D}$
$d(A)$	The diameter of the set A
$\mathrm{Exp}\,\mathscr{H}_G$ and $\mathrm{Exp}\,\mathscr{K}_\mu$	The Fock space
$G(\cdot)$	The spectral measure of a stationary discrete or generalized random field
$G_n \xrightarrow{v} G_0$	The vague convergence of the locally finite measures G_n to the locally finite measure G_0
$H_n(x)$	The Hermite polynomial of order n with leading coefficient 1
$\mathscr{H}$	The (real) Hilbert space of square-integrable random variables measurable with respect to the σ-algebra generated by the random variables of a previously defined Gaussian random field
$\mathscr{H}_1$	The smallest subspace of the Hilbert space $\mathscr{H}$ containing the elements of the underlying Gaussian field
$\mathscr{H}_{\le n}$	The smallest subspace of the Hilbert space $\mathscr{H}$ containing the polynomials of order less than or equal to n of the random variables in the underlying Gaussian field
$\mathscr{H}_n$	The orthogonal completion of the subspace $\mathscr{H}_{\le n-1}$ in the Hilbert space $\mathscr{H}_{\le n}$
$h_\gamma(\cdot)$	The kernel function of the Wiener–Itô integral appearing in the diagram formula and depending on the diagram γ
$\bar{\mathscr{H}}_G^n$	The space of functions which can be the kernel function of an n-fold Wiener–Itô integral with respect to a random spectral measure Z_G with spectral measure G
$\mathscr{H}_G^n$	The subspace of $\bar{\mathscr{H}}_G^n$ consisting of symmetric functions
$\hat{\mathscr{H}}_G^n$	The subspace of $\bar{\mathscr{H}}_G^n$ consisting of simple functions

$\hat{\mathscr{H}}_G^n$	The subspace of $\mathscr{H}_G^n$ consisting of simple (and symmetric) functions
$I_G(f_n)$	The normalized Wiener–Itô integral of the kernel function f_n of n variables with respect to the random spectral measure Z_G
$\bar{\mathscr{K}}_\mu^n$	The class of function which can be the kernel function of an n-fold Wiener–Itô integral with respect to a random orthogonal measure Z_μ
$\mathscr{K}_\mu^n$	The subspace of $\bar{\mathscr{K}}_\mu^n$ consisting of symmetric functions
$\hat{\bar{\mathscr{K}}}_\mu^n$	The set of simple functions appearing in the definition of n-fold Wiener–Itô integrals with respect to a random orthogonal measure Z_μ
$\mathscr{K}$	The Hilbert space of square integrable random variables measurable with respect to the σ-algebra generated by the random variables $Z_\mu(A)$ of a random orthogonal measure Z_μ
$\mathscr{K}_{\le n}$	The subspace of $\mathscr{K}$ generated by the polynomials of the random variables $Z_\mu(A)$ of the orthogonal random field Z_μ which have order less than or equal to n
$\mathscr{K}_n$	The orthogonal completion of the subspace $\mathscr{K}_{\le n-1}$ in the Hilbert space $\mathscr{K}_{\le n}$
$:P(\xi_1,\dots,\xi_n):$	The Wick polynomial corresponding to the polynomial $P(x_1,\dots,x_n)$ and Gaussian random vector $(\xi_1,\dots,\xi_n)$
$\mathscr{S}$	The class of test functions in the Schwartz space
$\mathscr{S}^c$	The class of complex number valued test functions in the Schwartz space
$\mathscr{S}'$	The Schwartz space of generalized functions
$S_{\nu-1}$	The ν-dimensional unit sphere
$\mathrm{Sym}\, f$	The symmetrization of the function f
T_m and T_t	The shift operator with parameter $m \in \mathbb{Z}_\nu$ and $t \in R^\nu$
$X(\varphi)$	The value of the generalized field $X(\cdot)$ at the test function φ
$Z_G(\cdot)$	The (Gaussian) random spectral measure corresponding to the spectral measure G
$Z(dx)$	The (Gaussian) random spectral measure whose spectral measure is $\frac{1}{2\pi}$ times the Lebesgue measure on $[-\pi,\pi)$
$Z_\mu(\cdot)$	The random orthogonal measure corresponding to the measure μ
$\mathbb{Z}_\nu$	The set of lattice points in the ν-dimensional space with integer coordinates
$\Gamma(n_1,\dots,n_m)$	The space of diagrams in the diagram formula
$\bar{\Gamma}$	The space of closed diagrams
$\lvert\gamma\rvert$	The number of edges in a diagram γ
$\mu_n \xrightarrow{w} \mu$	The weak convergence of the probability measures μ_n to the probability measure μ

$\xi_N \xrightarrow{\mathscr{D}} \xi_0$	The convergence of the random variables ξ_N to the random variable ξ_0 in distribution, i.e. the weak convergence of the distributions of ξ_N to the distribution of ξ_0
Π_n	The group of permutations of the set $\{1,\dots,n\}$
$\rho_p(\mu,\nu)$	The Prokhorov metric of the probability measures μ and ν
$\chi_A(\cdot)$	The indicator function of the set A.
$\tilde{\chi}_n(x)$	The Fourier transform of the indicator function of the unit cube $\prod_{p=1}^{\nu}[n^{(p)}, n^{(p)}+1)$, where $n=(n^{(1)},\dots,n^{(p)})$
$\ominus$	The orthogonal completion of a subspace of a Hilbert space
$\tilde{}$	Fourier transform
$*$	Convolutions
$\stackrel{\Delta}{=}$	Identity in distribution
$\Rightarrow$	Stochastic convergence
$\int'$	Wiener–Itô integral with respect to a random orthogonal measure
$[x]$	Integer part of a real number x

Chapter 1
On a Limit Problem

We begin with the formulation of a problem which is important both for probability theory and statistical physics. The multiple Wiener–Itô integral proved to be a very useful tool at the investigation of this problem.

Let us consider a set of random variables ξ_n, $n \in \mathbb{Z}_\nu$, where $\mathbb{Z}_\nu$ denotes the ν-dimensional integer lattice, and let us study their properties. Such a set of random variables will be called a (ν-dimensional) discrete random field. We shall be mainly interested in so-called stationary random fields. Let us recall their definition.

Definition of Discrete (Strictly) Stationary Random Fields. *A set of random variables ξ_n, $n \in \mathbb{Z}_\nu$, is called a (strictly) stationary discrete random field if*

$$(\xi_{n_1}, \dots, \xi_{n_k}) \stackrel{\Delta}{=} (\xi_{n_1+m}, \dots, \xi_{n_k+m})$$

for all $k = 1, 2, \dots$ and $n_1, \dots, n_k$, $m \in \mathbb{Z}_\nu$, where $\stackrel{\Delta}{=}$ denotes equality in distribution.

Let us also recall that a discrete random field ξ_n, $n \in \mathbb{Z}_\nu$, is called Gaussian if for every finite subset $\{n_1, \dots, n_k\} \subset \mathbb{Z}_\nu$ the random vector $(\xi_{n_1}, \dots, \xi_{n_k})$ is normally distributed.

Given a discrete random field ξ_n, $n \in \mathbb{Z}_\nu$, we define for all $N = 1, 2, \dots$ the new random fields

$$Z_n^N = A_N^{-1} \sum_{j \in B_n^N} \xi_j, \qquad N = 1, 2, \dots, \qquad n \in \mathbb{Z}_\nu, \tag{1.1}$$

where

$$B_n^N = \{j\colon\ j \in \mathbb{Z}_\nu, \quad n^{(i)} N \le j^{(i)} < (n^{(i)} + 1)N,\ i = 1, 2, \dots, \nu\},$$

and A_N, $A_N > 0$, is an appropriate norming constant. The superscript i denotes the i-th coordinate of a vector in this formula. We are interested in the question when

P. Major, *Multiple Wiener-Itô Integrals*, Lecture Notes in Mathematics 849, DOI 10.1007/978-3-319-02642-8_1,

the finite dimensional distributions of the random fields Z_n^N defined in (1.1) have a limit as $N \to \infty$. In particular, we would like to describe those random fields Z_n^*, $n \in \mathbb{Z}_\nu$, which appear as the limit of such random fields Z_n^N. This problem led to the introduction of the following notion.

Definition of Self-similar (Discrete) Random Fields. *A (discrete) random field ξ_n, $n \in \mathbb{Z}_\nu$, is called self-similar with self-similarity parameter α if the random fields Z_n^N defined in (1.1) with their help and the choice $A_N = N^\alpha$ satisfy the relation*

$$(\xi_{n_1}, \dots, \xi_{n_k}) \stackrel{\Delta}{=} (Z_{n_1}^N, \dots, Z_{n_k}^N) \tag{1.2}$$

for all $N = 1, 2, \dots$ and $n_1, \dots, n_k \in \mathbb{Z}_\nu$.

We are interested in the choice $A_N = N^\alpha$ with some $\alpha > 0$ in the definition of the random variables Z_n^N in (1.2), because under slight restrictions, relation (1.2) can be satisfied only with such norming constants A_N. A central problem both in statistical physics and in probability theory is the description of self-similar fields. We are interested in self-similar fields whose random variables have a finite second moment. This excludes the fields consisting of i.i.d. random variables with a non-Gaussian stable law.

The Gaussian self-similar random fields and their Gaussian range of attraction are fairly well known. Much less is known about the non-Gaussian case. The problem is hard, because the transformations of measures over $R^{\mathbb{Z}_\nu}$ induced by formula (1.1) have a very complicated structure. To get interesting results in some cases we shall define the so-called subordinated fields below. (More precisely, we define the fields subordinated to a stationary Gaussian field.) In case of subordinated fields the Wiener–Itô integral is a very useful tool for investigating the transformation defined in (1.1). In particular, it enables us to construct non-Gaussian self-similar fields and to prove non-trivial limit theorems. All known results are closely related to this technique.

Let X_n, $n \in \mathbb{Z}_\nu$, be a stationary Gaussian field. We define the shift transformations T_m, $m \in \mathbb{Z}_\nu$, over this field by the formula $T_m X_n = X_{n+m}$ for all $n, m \in \mathbb{Z}_\nu$. Let $\mathcal{H}$ denote the *real* Hilbert space consisting of the square integrable random variables measurable with respect to the σ-algebra $\mathcal{B} = \mathcal{B}(X_n,\ n \in \mathbb{Z}_\nu)$. The scalar product in $\mathcal{H}$ is defined as $(\xi, \eta) = E\xi\eta$, $\xi, \eta \in \mathcal{H}$. The shift transformations T_m, $m \in \mathbb{Z}_\nu$, can be extended to a group of unitary shift transformations over $\mathcal{H}$ in a natural way. Namely, if $\xi = f(X_{n_1}, \dots, X_{n_k})$ then we define $T_m\xi = f(X_{n_1+m}, \dots, X_{n_k+m})$. It can be seen that $\|\xi\| = \|T_m\xi\|$, and the above considered random variables ξ are dense in $\mathcal{H}$. (A more detailed discussion about the definition of shift operators and their properties will be given in Chap. 2 in a *Remark* after the formulation of Theorem 2C. Here we shall define the shift $T_m\xi$, $m \in \mathbb{Z}_\nu$, of all random variables ξ which are measurable with respect to the σ-algebra $\mathcal{B}(X_n,\ n \in Z_\nu)$, i.e. ξ does not have to be square integrable.) Hence T_m can be extended to the whole space $\mathcal{H}$ by L_2 continuity. It can be proved that the norm preserving transformations T_m, $m \in \mathbb{Z}_\nu$, constitute a unitary group in $\mathcal{H}$, i.e. $T_{n+m} = T_n T_m$ for all $n, m \in \mathbb{Z}_\nu$, and $T_0 = \text{Id}$. Now we introduce the following

Definition of Subordinated Random Fields. *Given a stationary Gaussian field X_n, $n \in \mathbb{Z}_\nu$, we define the Hilbert spaces $\mathcal{H}$ and the shift transformations T_m, $m \in \mathbb{Z}_\nu$, over $\mathcal{H}$ as before. A discrete stationary field ξ_n is called a random field subordinated to X_n if $\xi_n \in \mathcal{H}$, and $T_n \xi_m = \xi_{n+m}$ for all n, $m \in \mathbb{Z}_\nu$.*

We remark that ξ_0 determines the subordinated fields ξ_n completely, since $\xi_n = T_n \xi_0$. Later we give a more adequate description of subordinated fields by means of Wiener–Itô integrals. Before working out the details we formulate the continuous time version of the above notions and problems. In the continuous time case it is more natural to consider generalized random fields. To explain the idea behind such an approach we shortly explain a different but equivalent description of discrete random fields. We present them as an appropriate set of random variables indexed by the elements of a linear space. This shows some similarity with the generalized random fields to be defined later.

Let $\varphi_n(x)$, $n \in \mathbb{Z}_\nu$, $n = (n_1, \dots, n_\nu)$, denote the indicator function of the cube $[n_1 - \frac{1}{2}, n_1 + \frac{1}{2}) \times \cdots \times [n_\nu - \frac{1}{2}, n_\nu + \frac{1}{2})$, with center $n = (n_1, \dots, n_\nu)$ and with edges of length 1, i.e. let $\varphi_n(x) = 1$, $x = (x_1, \dots, x_\nu) \in R^\nu$, if $n_j - \frac{1}{2} \le x_j < n_j + \frac{1}{2}$ for all $1 \le j \le \nu$, and let $\varphi_n(x) = 0$ otherwise. Define the linear space Φ of functions on R^ν consisting of all finite linear combinations of the form $\sum c_j \varphi_{n_j}(x)$, $n_j \in \mathbb{Z}_\nu$, with the above defined functions $\varphi_n(x)$ and real coefficients c_j. Given a discrete random field ξ_n, $n \in \mathbb{Z}_\nu$, define the random variables $\xi(\varphi)$ for all $\varphi \in \Phi$ by the formula $\xi(\varphi) = \sum c_j \xi_{n_j}$ if $\varphi(x) = \sum c_j \varphi_{n_j}(x)$. In particular, $\xi(\varphi_n) = \xi_n$ for all $n \in \mathbb{Z}_\nu$. The identity $\xi(c_1\varphi + c_2\psi) = c_1\xi(\varphi) + c_2\xi(\psi)$ also holds for all $\varphi, \psi \in \Phi$ and real numbers c_1 and c_2.

Let us also define the function $\varphi^{(N,A_N)}(x) = \frac{1}{A_N}\varphi(\frac{x}{N})$ for all functions $\varphi \in \Phi$ and positive integers $N = 1, 2, \dots$, with some appropriately chosen constants $A_N > 0$. Observe that $\xi(\varphi_n^{(N,A_N)}) = Z_n^N$ with the random variable Z_n^N defined in (1.1). All previously introduced notions related to discrete random fields can be reformulated with the help of the set of random variables $\xi(\varphi)$, $\varphi \in \Phi$. Thus for instance the random field ξ_n, $n \in \mathbb{Z}_\nu$ is self-similar with self-similarity parameter α if and only if $\xi(\varphi^{(N,N^\alpha)}) \stackrel{\Delta}{=} \xi(\varphi)$ for all $\varphi \in \Phi$ and $N = 1, 2, \dots$. (To see why this statement holds observe that the distributions of two random vectors agree if and only if every linear combination of their coordinates have the same distribution. This follows from the fact that the characteristic function of a random vector determines its distribution.)

It will be useful to define the continuous time version of discrete random fields as generalized random fields. The generalized random fields will be defined as a set of random variables indexed by the elements of a linear space of functions. They show some similarity to the class of random variables $\xi(\varphi)$, $\varphi \in \Phi$, defined above. The main difference is that instead of the space Φ a different linear space is chosen for the parameter set of the random field. We shall choose the so-called Schwartz space for this role.

Let $\mathcal{S} = \mathcal{S}_\nu$ be the Schwartz space of (real valued) rapidly decreasing, smooth functions on R^ν. (See e.g. [16] for the definition of $\mathcal{S}_\nu$. I shall present a more detailed discussion about the definition of the space $\mathcal{S}$ together with the topology introduced in it in the adjustment to Chap. 1.) Generally one takes the space of complex valued, rapidly decreasing, smooth functions as the space $\mathcal{S}$, but we shall denote the space of *real valued*, rapidly decreasing, smooth functions by $\mathcal{S}$ if we do not say this otherwise. We shall omit the subscript ν if it leads to no ambiguity. Now we introduce the notion of generalized random fields.

Definition of Generalized Random Fields. *We say that the set of random variables $X(\varphi)$, $\varphi \in \mathcal{S}$, is a generalized random field over the Schwartz space $\mathcal{S}$ of rapidly decreasing, smooth functions if:*

(a) $X(a_1\varphi_1 + a_2\varphi_2) = a_1 X(\varphi_1) + a_2 X(\varphi_2)$ with probability 1 for all real numbers a_1 and a_2 and $\varphi_1 \in \mathcal{S}$, $\varphi_2 \in \mathcal{S}$. (The exceptional set of probability 0 where this identity does not hold may depend on a_1, a_2, φ_1 and φ_2.)
(b) $X(\varphi_n) \Rightarrow X(\varphi)$ stochastically if $\varphi_n \to \varphi$ in the topology of $\mathcal{S}$.

We also introduce the following definitions.

Definition of Stationarity and Gaussian Property of a Generalized Random Field and the Notion of Convergence of Generalized Random Fields in Distribution. *The generalized random field $X = \{X(\varphi), \varphi \in \mathcal{S}\}$ is stationary if $X(\varphi) \overset{\Delta}{=} X(T_t\varphi)$ for all $\varphi \in \mathcal{S}$ and $t \in R^\nu$, where $T_t\varphi(x) = \varphi(x-t)$. It is Gaussian if $X(\varphi)$ is a Gaussian random variable for all $\varphi \in \mathcal{S}$. The relation $X_n \xrightarrow{\mathcal{D}} X_0$ as $n \to \infty$ holds for a sequence of generalized random fields X_n, $n = 0, 1, 2, \dots$, if $X_n(\varphi) \xrightarrow{\mathcal{D}} X_0(\varphi)$ for all $\varphi \in \mathcal{S}$, where $\xrightarrow{\mathcal{D}}$ denotes convergence in distribution.*

Given a stationary generalized random field X and a function $A(t) > 0$, $t > 0$, on the set of positive real numbers we define the (stationary) random fields X_t^A for all $t > 0$ by the formula

$$X_t^A(\varphi) = X(\varphi_t^A), \quad \varphi \in \mathcal{S}, \qquad \text{where } \varphi_t^A(x) = A(t)^{-1}\varphi\left(\frac{x}{t}\right). \tag{1.3}$$

We are interested in the following

Question. *When does a generalized random field X^* exist such that $X_t^A \xrightarrow{\mathcal{D}} X^*$ as $t \to \infty$ (or as $t \to 0$)?*

In relation to this question we introduce the following

Definition of Self-similarity. *The stationary generalized random field X is self-similar with self-similarity parameter α if $X_t^A(\varphi) \overset{\Delta}{=} X(\varphi)$ for all $\varphi \in \mathcal{S}$ and $t > 0$ with the function $A(t) = t^\alpha$.*

To answer the above question one should first describe the generalized self-similar random fields.

We try to explain the motivation behind the above definitions. Given an ordinary random field $X(t)$, $t \in R^\nu$, and a topological space $\mathcal{E}$ consisting of functions over R^ν one can define the random variables $X(\varphi) = \int_{R^\nu} \varphi(t)X(t)\,dt$, $\varphi \in \mathcal{E}$. Some difficulty may arise when defining this integral, but it can be overcome in all interesting cases. If the space $\mathcal{E}$ is rich enough, and this is the case if $\mathcal{E} = \mathcal{S}$, then the integrals $X(\varphi)$, $\varphi \in \mathscr{E}$, determine the random process $X(t)$. The set of random variables $X(\varphi)$, $\varphi \in \mathscr{E}$, is a generalized random field in all nice cases. On the other hand, there are generalized random fields which cannot be obtained by integrating ordinary random fields. In particular, the generalized self-similar random fields we shall construct later cannot be interpreted through ordinary fields. The above definitions of various properties of generalized fields are fairly natural, considering what these definitions mean for generalized random fields obtained by integrating ordinary fields.

The investigation of generalized random fields is simpler than that of ordinary discrete random fields, because in the continuous case more symmetry is available. Moreover, in the study or construction of discrete random fields generalized random fields may play a useful role. To understand this let us remark that if we have a generalized random field $X(\varphi)$, $\varphi \in \mathscr{S}$, and we can extend the space $\mathscr{S}$ containing the test function φ to such a larger linear space $\mathscr{T}$ for which $\Phi \subset \mathscr{T}$ with the above introduced linear space Φ, then we can define the discrete random field $X(\varphi)$, $\varphi \in \Phi$, by a restriction of the space of test functions of the generalized random field $X(\varphi)$, $\varphi \in \mathscr{T}$. This random field can be considered as the discretization of the original generalized random field $X(\varphi)$, $\varphi \in \mathscr{S}$.

We finish this chapter by defining the generalized subordinated random fields. Then we shall explain the basic results about the Schwartz space $\mathscr{S}$ and generalized functions in a separate sub chapter.

Let $X(\varphi)$, $\varphi \in \mathscr{S}$, be a generalized stationary Gaussian random field. The formula $T_t X(\varphi)) = X(T_t\varphi)$, $T_t\varphi(x) = \varphi(x-t)$, defines the shift transformation for all $t \in R^\nu$. Let $\mathscr{H}$ denote the real Hilbert space consisting of the $\mathscr{B} = \mathscr{B}(X(\varphi),\ \varphi \in \mathscr{S})$ measurable random variables with finite second moment. The shift transformation can be extended to a group of unitary transformations over $\mathscr{H}$ similarly to the discrete case. This will be explained in more detail in the next chapter.

Definition of Generalized Random Fields Subordinated to a Generalized Stationary Gaussian Random Field. *Given a generalized stationary Gaussian random field $X(\varphi)$, $\varphi \in \mathscr{S}$, we define the Hilbert space $\mathscr{H}$ and the shift transformations T_t, $t \in R^\nu$, over $\mathscr{H}$ as above. A generalized stationary random field $\xi(\varphi)$, $\varphi \in \mathscr{S}$, is subordinated to the field $X(\varphi)$, $\varphi \in \mathscr{S}$, if $\xi(\varphi) \in \mathscr{H}$ and $T_t\xi(\varphi) = \xi(T_t\varphi)$ for all $\varphi \in \mathscr{S}$ and $t \in R^\nu$, and $E[\xi\varphi_n) - \xi(\varphi)]^2 \to 0$ if $\varphi_n \to \varphi$ in the topology of $\mathscr{S}$.*

1.1 A Brief Overview About Some Results on Generalized Functions

Let us first describe the Schwartz spaces $\mathscr{S}$ and $\mathscr{S}^c$ in more detail. The space $\mathscr{S}^c = (\mathscr{S}_\nu)^c$ consists of those complex valued functions of ν variables which decrease at infinity, together with their derivatives, faster than any polynomial degree. More explicitly, $\varphi \in \mathscr{S}^c$ for a complex valued function φ of ν variables if

$$\left| x_1^{k_1} \cdots x_\nu^{k_\nu} \frac{\partial^{q_1+\cdots+q_\nu}}{\partial x_1^{q_1} \ldots \partial x_\nu^{q_\nu}} \varphi(x_1, \ldots, x_\nu) \right| \le C(k_1, \ldots, k_\nu, q_1, \ldots, q_\nu)$$

for all point $x = (x_1, \ldots, x_\nu) \in R^\nu$ and vectors $(k_1, \ldots, k_\nu)$, $(q_1, \ldots, q_\nu)$ with non-negative integer coordinates with some constant $C(k_1, \ldots, k_\nu, q_1, \ldots, q_\nu)$ which may depend on the function φ. This formula can be written in a more concise form as

$$|x^k D^q \varphi(x)| \le C(k,q) \quad \text{with } k = (k_1, \ldots, k_\nu) \text{ and } q = (q_1, \ldots, q_\nu),$$

where $x = (x_1, \ldots, x_\nu)$, $x^k = x_1^{k_1} \cdots x_\nu^{k_\nu}$ and $D^q = \frac{\partial^{q_1+\cdots+q_\nu}}{\partial x_1^{q_1} \ldots \partial x_\nu^{q_\nu}}$. The elements of the space $\mathcal{S}$ are defined similarly, with the only difference that they are real valued functions.

To define the spaces $\mathcal{S}$ and $\mathcal{S}^c$ we still have to define the convergence in them. We say that a sequence of functions $\varphi_n \in \mathcal{S}^c$ (or $\varphi_n \in \mathcal{S}$) converges to a function φ if

$$\lim_{n\to\infty} \sup_{x \in R^\nu} (1 + |x|^2)^k |D^q \varphi_n(x) - D^q \varphi(x)| = 0.$$

for all $k = 1, 2, \ldots$ and $q = (q_1, \ldots, q_\nu)$. It can be seen that the limit function φ is also in the space $\mathcal{S}^c$ (or in the space $\mathcal{S}$).

A nice topology can be introduced in the space $\mathcal{S}^c$ (or $\mathcal{S}$) which induces the above convergence. The following topology is an appropriate choice. Let a basis of neighbourhoods of the origin consist of the sets

$$U(k, q, \varepsilon) = \left\{ \varphi \colon \max_x (1 + |x|^2)^k |D^q \varphi(x)| < \varepsilon \right\}$$

with $k = 0, 1, 2, \ldots$, $q = (q_1, \ldots, q_\nu)$ with non-negative integer coordinates and $\varepsilon > 0$, where $|x|^2 = x_1^2 + \cdots + x_\nu^2$. A basis of neighbourhoods of an arbitrary function $\varphi \in \mathcal{S}^c$ (or $\varphi \in \mathcal{S}$) consists of sets of the form $\varphi + U(k, q, \varepsilon)$, where the class of sets $U(k, q, \varepsilon)$ is a basis of neighbourhood of the origin. The fact that the convergence in $\mathcal{S}$ has such a representation, (and a similar result holds in some other spaces studied in the theory of generalized functions) has a great importance in the theory of generalized functions. We also have exploited this fact in Chap. 6 of this Lecture Note. Topological spaces with such a topology are called countably normed spaces.

The space of generalized functions $\mathcal{S}'$ consists of the *continuous* linear maps $F \colon \mathcal{S} \to C$ or $F \colon \mathcal{S}^c \to C$, where C denotes the linear space of complex numbers. (In the study of the space $\mathcal{S}'$ we omit the upper index c, i.e. we do not indicate whether we are working in real or complex space when this causes no problem.) We shall write the map $F(\varphi)$, $F \in \mathcal{S}'$ and $\varphi \in \mathcal{S}$ (or $\varphi \in \mathcal{S}^c$) in the form (F, φ).

We can define generalized functions $F \in \mathcal{S}'$ by the formula

$$(F, \varphi) = \int \overline{f(x)}\varphi(x)\,dx \quad \text{for all } \varphi \in \mathcal{S} \quad \text{or } \varphi \in \mathcal{S}^c$$

with a function f such that $\int (1+|x|^2)^{-p}|f(x)|\,dx < \infty$ with some $p \geq 0$. (The upper script $\bar{}$ denotes complex conjugate in the sequel.) Such functionals are called regular. There are also non-regular functionals in the space $\mathcal{S}'$. An example for them is the δ-function defined by the formula $(\delta, \varphi) = \varphi(0)$. There is a good description of the generalized functions $F \in \mathcal{S}'$, (see the book I.M. Gelfand and G.E. Shilov: Generalized functions, Vol. 2, Chaps 2, 4), but we do not need this result, hence we do not discuss it here. Another important question in this field not discussed in the present note is about the interpretation of a usual function as a generalized function in the case when it does not define a regular function in $\mathcal{S}'$ because of its strong singularity in some points. In such cases some regularization can be applied. It is an important problem in the theory of generalized functions to define the appropriate generalized functions in such cases, but it does not appear in the study of the problems in this work.

The derivative and the Fourier transform of generalized functions are also defined, and they play an important role in some investigations. In the definition of these notions for generalized functions we want to preserve the old definition if nice regular functionals are considered for which these notions were already defined in classical analysis. Such considerations lead to the definition $(\frac{\partial_j}{\partial x_j}F, \varphi) = -(F, \frac{\partial \varphi}{\partial x_j})$ of the derivative of generalized functions. We do not discuss this definition in more detail, because here we do not work with the derivatives of generalized functions.

The Fourier transform of generalized functions in S' appears in our discussion, although only in an implicit form. The Bochner–Schwartz theorem discussed in Chap. 3 actually deals with the Fourier transform of generalized functions. Hence the definition of Fourier transform will be given in more detail.

We shall define the Fourier transform of a generalized function by means of a natural extension of the Parseval formula, more explicitly of a simplified version of it, where the same identity

$$\int_{R^\nu} \overline{f(x)}g(x)\,dx = \frac{1}{(2\pi)^\nu}\int_{R^\nu} \overline{\tilde{f}(u)}\tilde{g}(u)\,du$$

is formulated with $\tilde{f}(u) = \int_{R^\nu} e^{i(u,x)} f(x)\,dx$ and $\tilde{g}(u) = \int_{R^\nu} e^{i(u,x)} g(x)\,dx$. But now we consider a pair of functions (f, g) with different properties. We demand that f should be an integrable function, and $g \in \mathcal{S}^c$. (In the original version of the Parseval formula both f and g are L_2 functions.)

The proof of this identity is simple. Indeed, since the function $g \in \mathcal{S}^c$ can be calculated as the inverse Fourier transform of its Fourier transform $\tilde{g} \in \mathcal{S}^c$, i.e. $g(x) = \frac{1}{(2\pi)^\nu}\int e^{-i(u,x)}\tilde{g}(u)\,du$, we can write

$$\int \overline{f(x)} g(x)\, dx = \int \overline{f(x)} \left[\frac{1}{(2\pi)^\nu} \int e^{-i(u,x)} \tilde{g}(u)\, du \right] dx$$
$$= \int \tilde{g}(u) \left[\frac{1}{(2\pi)^\nu} \int \overline{e^{i(u,x)} f(x)\, dx} \right] du$$
$$= \frac{1}{(2\pi)^\nu} \int \overline{\tilde{f}(u)} \tilde{g}(u)\, du.$$

Let us also remark that the Fourier transform $f \to \tilde{f}$ is a bicontinuous map from $\mathcal{S}^c$ to $\mathcal{S}^c$. (This means that this transformation is invertible, and both the Fourier transform and its inverse are continuous maps from $\mathcal{S}^c$ to $\mathcal{S}^c$.) (The restriction of the Fourier transform to the space $\mathcal{S}$ of real valued functions is a bicontinuous map from $\mathcal{S}$ to the subspace of $\mathcal{S}^c$ consisting of those functions $f \in \mathcal{S}^c$ for which $f(-x) = \overline{f(x)}$ for all $x \in R^\nu$.)

The above results make natural the following definition of the Fourier transform $\tilde{F}$ of a generalized function $F \in \mathcal{S}'$.

$$(\tilde{F}, \tilde{\varphi}) = (2\pi)^\nu (F, \varphi) \quad \text{for all } \varphi \in \mathcal{S}^c.$$

Indeed, if $F \in \mathcal{S}'$ then $\tilde{F}$ is also a continuous linear map on $\mathcal{S}^c$, i.e. it is also an element of $\mathcal{S}'$. Besides, the above proved version of the Parseval formula implies that if we consider an integrable function f on R^ν both as a usual function and as a (regular) generalized function, its Fourier transform agrees in the two cases.

There are other classes of test functions and spaces of generalized functions studied in the literature. The most popular among them is the space $\mathcal{D}$ of infinitely many times differentiable functions with compact support and its dual space $\mathcal{D}'$, the space of continuous linear transformations on the space $\mathcal{D}$. (These spaces are generally denoted by $\mathcal{D}$ and $\mathcal{D}'$ in the literature, although just the book [16] that we use as our main reference in this subject applies the notation $\mathcal{K}$ and $\mathcal{K}'$ for them.) We shall discuss this space only very briefly.

The space $\mathcal{D}$ consists of the infinitely many times differentiable functions with compact support. Thus it is a subspace of $\mathcal{S}$. A sequence $\varphi_n \in \mathcal{D}$, $n = 1, 2, \dots$, converges to a function φ, if there is a compact set $A \subset R^\nu$ which is the support of all these functions φ_n, and the functions φ_n together with all their derivatives converge uniformly to the function φ and to its corresponding derivatives. It is not difficult to see that also $\varphi \in \mathcal{D}$, and if the functions φ_n converge to φ in the space $\mathcal{D}$, then they also converge to φ in the space $\mathcal{S}$. Moreover, $\mathcal{D}$ is an everywhere dense subspace of $\mathcal{S}$. The space $\mathcal{D}'$ consists of the continuous linear functionals in $\mathcal{D}$.

The results describing the behaviour of $\mathcal{D}$ and $\mathcal{D}'$ are very similar to those describing the behaviour of $\mathcal{S}$ and $\mathcal{S}'$. There is one difference that deserves some attention. The Fourier transforms of the functions in $\mathcal{D}$ may not belong to $\mathcal{D}$. The class of these Fourier transforms can be described by means of some results in complex analysis. A topological space $\mathcal{Z}$ can be defined on the set of Fourier transforms of the functions from the space $\mathcal{D}$. If we want to apply Fourier analysis in the space $\mathcal{D}$, then we also have to study this space $\mathcal{Z}$ and its dual space $\mathcal{Z}'$. I omit the details.

Chapter 2
Wick Polynomials

In this chapter we consider the so-called Wick polynomials, a multi-dimensional generalization of Hermite polynomials. They are closely related to multiple Wiener–Itô integrals.

Let X_t, $t \in T$, be a set of jointly Gaussian random variables indexed by a parameter set T. Let $EX_t = 0$ for all $t \in T$. We define the real Hilbert spaces $\mathcal{H}_1$ and $\mathcal{H}$ in the following way: A square integrable random variable is in $\mathcal{H}$ if and only if it is measurable with respect to the σ-algebra $\mathcal{B} = \mathcal{B}(X_t,\ t \in T)$, and the scalar product in $\mathcal{H}$ is defined as $(\xi, \eta) = E\xi\eta$, $\xi, \eta \in \mathcal{H}$. The Hilbert space $\mathcal{H}_1 \subset \mathcal{H}$ is the subspace of $\mathcal{H}$ generated by the finite linear combinations $\sum c_j X_{t_j}$, $t_j \in T$. We consider only such sets of Gaussian random variables X_t for which $\mathcal{H}_1$ is separable. Otherwise X_t, $t \in T$, can be arbitrary, but the most interesting case for us is when $T = \mathcal{S}_\nu$ or $\mathbb{Z}_\nu$, and X_t, $t \in T$, is a stationary Gaussian field.

Let $Y_1, Y_2, \dots$ be an orthonormal basis in $\mathcal{H}_1$. The uncorrelated random variables $Y_1, Y_2, \dots$ are independent, since they are (jointly) Gaussian. Moreover,

$$\mathcal{B}(Y_1, Y_2, \dots) = \mathcal{B}(X_t,\ t \in T).$$

Let $H_n(x)$ denote the n-th Hermite polynomial with leading coefficient 1, i.e. let $H_n(x) = (-1)^n e^{x^2/2} \frac{d^n}{dx^n}(e^{-x^2/2})$. We recall the following results from analysis and measure theory.

Theorem 2A. *The Hermite polynomials $H_n(x)$, $n = 0, 1, 2, \dots$, form a complete orthogonal system in $L_2\left(R, \mathcal{B}, \frac{1}{\sqrt{2\pi}} e^{-x^2/2}\, dx\right)$. (Here $\mathcal{B}$ denotes the Borel σ-algebra on the real line.)*

Let $(X_j, \mathcal{X}_j, \mu_j)$, $j = 1, 2, \dots$, be countably many independent copies of a probability space $(X, \mathcal{X}, \mu)$. (We denote the points of X_j by x_j.) Let $(X^\infty, \mathcal{X}^\infty, \mu^\infty) = \prod_{j=1}^{\infty} (X_j, \mathcal{X}_j, \mu_j)$. With such a notation the following result holds.

P. Major, *Multiple Wiener-Itô Integrals*, Lecture Notes in Mathematics 849, DOI 10.1007/978-3-319-02642-8_2,

Theorem 2B. *Let $\varphi_0, \varphi_1, \dots,$ $\varphi_0(x) \equiv 1$, be a complete orthonormal system in the Hilbert space $L_2(X, \mathscr{X}, \mu)$. Then the functions $\prod_{j=1}^{\infty} \varphi_{k_j}(x_j)$, where only finitely many indices k_j differ from 0, form a complete orthonormal basis in $L_2(X^\infty, \mathscr{X}^\infty, \mu^\infty)$.*

Theorem 2C. *Let $Y_1, Y_2, \dots$ be random variables on a probability space $(\Omega, \mathscr{A}, P)$ taking values in a measurable space $(X, \mathscr{X})$. Let ξ be a real valued random variable measurable with respect to the σ-algebra $\mathscr{B}(Y_1, Y_2, \dots)$, and let $(X^\infty, \mathscr{X}^\infty)$ denote the infinite product $(X \times X \times \cdots, \mathscr{X} \times \mathscr{X} \times \cdots)$ of the space $(X, \mathscr{X})$ with itself. Then there exists a real valued, measurable function f on the space $(X^\infty, \mathscr{X}^\infty)$ such that $\xi = f(Y_1, Y_2, \dots)$.*

Remark. Let us have a stationary random field $X_n(\omega)$, $n \in \mathbb{Z}_\nu$. Theorem 2C enables us to extend the shift transformation T_m, defined as $T_m X_n(\omega) = X_{n+m}(\omega)$, $n, m \in \mathbb{Z}_\nu$, for all random variables $\xi(\omega)$, measurable with respect to the σ-algebra $\mathscr{B}(X_n(\omega), n \in \mathbb{Z}_\nu)$. Indeed, by Theorem 2C we can write $\xi(\omega) = f(X_n(\omega), n \in \mathbb{Z}_\nu)$, and define $T_m\xi(\omega) = f(X_{n+m}(\omega), n \in \mathbb{Z}_\nu)$. We still have to understand, that although the function f is not unique in the representation of the random variable $\xi(\omega)$, the above definition of $T_m\xi(\omega)$ is meaningful. To see this we have to observe that if $f_1(X_n(\omega), n \in \mathbb{Z}_\nu) = f_2(X_n(\omega), n \in \mathbb{Z}_\nu)$ for two functions f_1 and f_2 with probability 1, then also $f_1(X_{n+m}(\omega), n \in \mathbb{Z}_\nu) = f_2(X_{n+m}(\omega), n \in \mathbb{Z}_\nu)$ with probability 1 because of the stationarity of the random field $X_n(\omega)$, $n \in \mathbb{Z}_\nu$. Let us also observe that $\xi(\omega) \stackrel{\Delta}{=} T_m\xi(\omega)$ for all $m \in \mathbb{Z}_\nu$. Besides, T_m is a linear operator on the linear space of random variables, measurable with respect to the σ-algebras $\mathscr{B}(X_n, n \in \mathbb{Z}_\nu)$. If we restrict it to the space of square integrable random variables, then T_m is a unitary operator, and the operators T_m, $m \in \mathbb{Z}_\nu$, constitute a unitary group.

Let a stationary generalized random field $X = \{X(\varphi), \varphi \in \mathscr{S}\}$ be given. The shift $T_t\xi$ of a random variable ξ, measurable with respect to the σ-algebra $\mathscr{B}(X(\varphi), \varphi \in \mathscr{S})$ can be defined for all $t \in R^\nu$ similarly to the discrete case with the help of Theorem 2C and the following result. If $\xi \in \mathscr{B}(X(\varphi), \varphi \in \mathscr{S})$ for a random variable ξ, then there exists such a countable subset $\{\varphi_1, \varphi_2, \dots\} \subset \mathscr{S}$ (depending on the random variable ξ) for which ξ is $\mathscr{B}(X(\varphi_1), X(\varphi_2), \dots)$ measurable. (We write $\xi(\omega) = f(X(\varphi_1)(\omega), X(\varphi_2)(\omega), \dots)$ with appropriate functions f, and $\varphi_1 \in \mathscr{S}$, $\varphi_2 \in \mathscr{S}$,..., and define the shift $T_t\xi$ as $T_t\xi(\omega) = f(X(T_t\varphi_1)(\omega), X(T_t\varphi_2)(\omega), \dots)$, where $T_t\varphi(x) = \varphi(x - t)$ for $\varphi \in \mathscr{S}$.) The transformations T_t, $t \in R^\nu$, are linear operators over the space of random variables measurable with respect to the σ-algebra $\mathscr{B}(X(\varphi), \varphi \in \mathscr{S})$ with similar properties as their discrete counterpart.

Theorems 2A–2C have the following important consequence.

Theorem 2.1. *Let $Y_1, Y_2, \dots$ be an orthonormal basis in the Hilbert space $\mathscr{H}_1$ defined above with the help of a set of Gaussian random variables X_t, $t \in T$. Then the set of all possible finite products $H_{j_1}(Y_{l_1}) \cdots H_{j_k}(Y_{l_k})$ is a complete orthogonal* *system in the Hilbert space $\mathcal{H}$ defined above. (Here $H_j(\cdot)$ denotes the j-th Hermite polynomial.)*

Proof of Theorem 2.1. By Theorems 2A and 2B the set of all possible products $\prod_{j=1}^{\infty} H_{k_j}(x_j)$, where only finitely many indices k_j differ from 0, is a complete orthonormal system in $L_2\left(R^\infty, \mathcal{B}^\infty, \prod_{j=1}^{\infty} \frac{e^{-x_j^2/2}}{\sqrt{2\pi}}\,dx_j\right)$. Since $\mathcal{B}(X_t,\ t \in T) = \mathcal{B}(Y_1, Y_2, \dots)$, Theorem 2C implies that the mapping $f(x_1, x_2, \dots,) \to f(Y_1, Y_2, \dots)$ is a unitary transformation from $L_2\left(R^\infty, \mathcal{B}^\infty, \prod_{j=1}^{\infty} \frac{e^{-x_j^2/2}}{\sqrt{2\pi}}\,dx_j\right)$ to $\mathcal{H}$. (We call a transformation from a Hilbert space to another Hilbert space unitary if it is norm preserving and invertible.) Since the image of a complete orthogonal system under a unitary transformation is again a complete orthogonal system, Theorem 2.1 is proved. □

Let $\mathcal{H}_{\le n} \subset \mathcal{H}$, $n = 1, 2, \dots,$ (with the previously introduced Hilbert space $\mathcal{H}$) denote the Hilbert space which is the closure of the linear space consisting of the elements $P_n(X_{t_1}, \dots, X_{t_m})$, where P_n runs through all polynomials of degree less than or equal to n, and the integer m and indices $t_1, \dots, t_m \in T$ are arbitrary. Let $\mathcal{H}_0 = \mathcal{H}_{\le 0}$ consist of the constant functions, and let $\mathcal{H}_n = \mathcal{H}_{\le n} \ominus \mathcal{H}_{\le n-1}$, $n = 1, 2, \dots,$ where $\ominus$ denotes orthogonal completion. It is clear that the Hilbert space $\mathcal{H}_1$ given in this definition agrees with the previously defined Hilbert space $\mathcal{H}_1$. If $\xi_1, \dots, \xi_m \in \mathcal{H}_1$, and $P_n(x_1, \dots, x_m)$ is a polynomial of degree n, then $P_n(\xi_1, \dots, \xi_m) \in \mathcal{H}_{\le n}$. Hence Theorem 2.1 implies that

$$\mathcal{H} = \mathcal{H}_0 + \mathcal{H}_1 + \mathcal{H}_2 + \cdots, \tag{2.1}$$

where $+$ denotes direct sum. Now we introduce the following

Definition of Wick Polynomials. *Given a polynomial $P(x_1, \dots, x_m)$ of degree n and a set of (jointly) Gaussian random variables $\xi_1, \dots, \xi_m \in \mathcal{H}_1$, the Wick polynomial $:P(\xi_1, \dots, \xi_m):$ is the orthogonal projection of the random variable $P(\xi_1, \dots, \xi_m)$ to the above defined subspace $\mathcal{H}_n$ of the Hilbert space $\mathcal{H}$.*

It is clear that Wick polynomials of different degree are orthogonal. Given some $\xi_1, \dots, \xi_m \in \mathcal{H}_1$ define the subspaces $\mathcal{H}_{\le n}(\xi_1, \dots, \xi_m) \subset \mathcal{H}_{\le n}$, $n = 1, 2, \dots,$ as the set of all polynomials of the random variables $\xi_1, \dots, \xi_m$ with degree less than or equal to n. Let $\mathcal{H}_{\le 0}(\xi_1, \dots, \xi_m) = \mathcal{H}_0(\xi_1, \dots, \xi_m) = \mathcal{H}_0$, and $\mathcal{H}_n(\xi_1, \dots, \xi_m) = \mathcal{H}_{\le n}(\xi_1, \dots, \xi_m) \ominus \mathcal{H}_{\le n-1}(\xi_1, \dots, \xi_m)$. With the help of this notation we formulate the following

Proposition 2.2. *Let $P(x_1, \dots, x_m)$ be a polynomial of degree n. Then the random polynomial $:P(\xi_1, \dots, \xi_m):$ equals the orthogonal projection of $P(\xi_1, \dots, \xi_m)$ to $\mathcal{H}_n(\xi_1, \dots, \xi_m)$.*

Proof of Proposition 2.2. Let $:\bar{P}(\xi_1,\dots,\xi_m):$ denote the projection of the random polynomial $P(\xi_1,\dots,\xi_m)$ to $\mathcal{H}_n(\xi_1,\dots,\xi_m)$. Obviously

$$P(\xi_1,\dots,\xi_m) - :\bar{P}(\xi_1,\dots,\xi_m): \in \mathcal{H}_{\le n-1}(\xi_1,\dots,\xi_m) \subseteq \mathcal{H}_{\le n-1}.$$

Hence in order to prove Proposition 2.2 it is enough to show that for all $\eta \in \mathcal{H}_{\le n-1}$

$$E:\bar{P}(\xi_1,\dots,\xi_m):\eta = 0, \tag{2.2}$$

since this means that $:\bar{P}(\xi_1,\dots,\xi_m):$ is the orthogonal projection of $P(\xi_1,\dots,\xi_m) \in \mathcal{H}_{\le n}$ to $\mathcal{H}_{\le n-1}$.

Let $\varepsilon_1,\varepsilon_2,\dots$ be an orthonormal system in $\mathcal{H}_1$, also orthonormal to $\xi_1,\dots,\xi_m$, and such that $\xi_1,\dots,\xi_m,\varepsilon_1,\varepsilon_2,\dots$ form a basis in $\mathcal{H}_1$. If $\eta = \prod\limits_{i=1}^{m} \xi_i^{l_i} \prod\limits_{j=1}^{\infty} \varepsilon_j^{k_j}$ with such exponents l_i and k_j that $\sum l_i + \sum k_j \le n-1$, then (2.2) holds for this random variable η because of the independence of the random variables ξ_i and ε_j. Since the linear combinations of such η are dense in $\mathcal{H}_{\le n-1}$, formula (2.2) and Proposition 2.2 are proved. □

Corollary 2.3. *Let $\xi_1,\dots,\xi_m$ be an orthonormal system in $\mathcal{H}_1$, and let*

$$P(x_1,\dots,x_m) = \sum c_{j_1,\dots,j_m} x^{j_1} \cdots x_m^{j_m}$$

be a homogeneous polynomial, i.e. let $j_1 + \cdots j_m = n$ with some fixed number n for all sets $(j_1,\dots,j_m)$ appearing in this summation. Then

$$:P(\xi_1,\dots,\xi_m): = \sum c_{j_1,\dots,j_m} H_{j_1}(\xi_1)\cdots H_{j_m}(\xi_m).$$

In particular,

$$:\xi^n: = H_n(\xi) \quad \textit{if } \xi \in \mathcal{H}_1, \textit{ and } E\xi^2 = 1.$$

Remark. Although we have defined the Wick polynomial (of degree n) for all polynomials $P(\xi_1,\dots,\xi_m)$ of degree n, we could have restricted our attention only to homogeneous polynomials of degree n, since the contribution of each terms $c(j_1,\dots j_m)\xi_1^{l_1}\cdots\xi_m^{l_m}$ of the polynomial $P(\xi_1,\dots,\xi_m)$ such that $l_1+\cdots+l_m < n$ has a zero contribution in the definition of the Wick polynomial $:P(\xi_1,\dots,\xi_m):$.

Proof of Corollary 2.3. Let the degree of the polynomial P be n. Then

$$P(\xi_1,\dots,\xi_m) - \sum c_{j_1,\dots,j_m} H_{j_1}(\xi_1)\cdots H_{j_m}(\xi_m) \in \mathcal{H}_{\le n-1}(\xi_1,\dots,\xi_m), \tag{2.3}$$

since $P(\xi_1,\dots,x_m)-\sum c_{j_1,\dots,j_m}H_{j_1}(\xi_1)\cdots H_{j_m}(\xi_m)$ is a polynomial whose degree is less than n. Let $\eta=\xi_1^{l_1}\cdots\xi_m^{l_m}$, $\sum\limits_{i=1}^m l_i\le n-1$. Then

$$E\eta H_{j_1}(\xi_1)\cdots H_{j_m}(\xi_m)=\prod_{i=1}^m E\xi_i^{l_i}H_{j_i}(\xi_i)=0,$$

since $l_i<j_i$ for at least one index i. Therefore

$$E\eta\sum c_{j_1,\dots,j_m}H_{j_1}(\xi_1)\cdots H_{j_m}(\xi_m)=0. \tag{2.4}$$

Since every element of $\mathcal{H}_{\le n-1}(\xi_1,\dots,\xi_m)$ can be written as the sum of such elements η, relation (2.4) holds for all $\eta\in\mathcal{H}_{\le n-1}(\xi_1,\dots,\xi_m)$. Relations (2.3) and (2.4) imply Corollary 2.3. □

The following statement is a simple consequence of the previous results.

Corollary 2.4. *Let $\xi_1,\xi_2,\dots$ be an orthonormal basis in $\mathcal{H}_1$. Then the random variables $H_{j_1}(\xi_1)\cdots H_{j_k}(\xi_k)$, $k=1,2,\dots$, $j_1+\dots+j_k=n$, form a complete orthogonal basis in $\mathcal{H}_n$.*

Proof of Corollary 2.4. It follows from Corollary 2.3 that

$$H_{j_1}(\xi_1)\cdots H_{j_k}(\xi_k)=:\xi_1^{j_1}\cdots\xi_k^{j_k}:\in\mathcal{H}_n\quad\text{for all } k=1,2,\dots$$

if $j_1+\dots+j_k=n$. These random variables are orthogonal, and all Wick polynomials $:P(\xi_1,\dots,\xi_m):$ of degree n of the random variables $\xi_1,\xi_2,\dots$ can be represented as the linear combination of such terms. Since these Wick polynomials are dense in $\mathcal{H}_n$, this implies Corollary 2.4. □

The arguments of this chapter exploited heavily some properties of Gaussian random variables. Namely, they exploited that the linear combinations of Gaussian random variables are again Gaussian, and in Gaussian case orthogonality implies independence. This means in particular, that the rotation of a standard normal vector leaves its distribution invariant. We finish this chapter with an observation based on these facts. This may illuminate the content of formula (2.1) from another point of view. We shall not use the results of the subsequent considerations in the rest of this work.

Let U be a unitary transformation over $\mathcal{H}_1$. It can be extended to a unitary transformation $\mathcal{U}$ over $\mathcal{H}$ in a natural way. Fix an orthonormal basis $\xi_1,\xi_2,\dots$ in $\mathcal{H}_1$, and define $\mathcal{U}1=1$, $\mathcal{U}\xi_{j_1}^{l_1}\cdots\xi_{j_k}^{l_k}=(U\xi_{j_1})^{l_1}\cdots(U\xi_{j_k})^{l_k}$. This transformation can be extended to a linear transformation $\mathcal{U}$ over $\mathcal{H}$ in a unique way. The transformation $\mathcal{U}$ is norm preserving, since the joint distributions of $(\xi_{j_1},\xi_{j_2},\dots)$ and $(U\xi_{j_1},U\xi_{j_2},\dots)$ coincide. Moreover, it is unitary, since $U\xi_1,U\xi_2,\dots$ is an orthonormal basis in $\mathcal{H}_1$. It is not difficult to see that if $P(x_1,\dots,x_m)$ is an arbitrary polynomial, and $\eta_1, \eta_2 \dots, \eta_m \in \mathscr{H}_1$, then $\mathscr{U} P(\eta_1, \dots, \eta_m) = P(U\eta_1, \dots, U\eta_m)$. This relation means in particular that the transformation $\mathscr{U}$ does not depend on the choice of the basis in $\mathscr{H}_1$. If the transformations $\mathscr{U}_1$ and $\mathscr{U}_2$ correspond to two unitary transformations U_1 and U_2 on $\mathscr{H}_1$, then the transformation $\mathscr{U}_1\mathscr{U}_2$ corresponds to U_1U_2. The subspaces $\mathscr{H}_{\le n}$ and therefore the subspaces $\mathscr{H}_n$ remain invariant under the transformations $\mathscr{U}$.

The shift transformations of a stationary Gaussian field, and their extensions to $\mathscr{H}$ are the most interesting examples for such unitary transformations U and $\mathscr{U}$. In the terminology of group representations the above facts can be formulated in the following way: The mapping $U \to \mathscr{U}$ is a group representation of $U(\mathscr{H}_1)$ over $\mathscr{H}$, where $U(\mathscr{H}_1)$ denotes the group of unitary transformations over $\mathscr{H}_1$. Formula (2.1) gives a decomposition of $\mathscr{H}$ into orthogonal invariant subspaces of this representation.

Chapter 3
Random Spectral Measures

Some standard theorems of probability theory state that the correlation function of a stationary random field can be expressed as the Fourier transform of a so-called spectral measure. In this chapter we construct a random measure with the help of these results, and express the random field itself as the Fourier transform of this random measure in some sense. We restrict ourselves to the Gaussian case, although most of the results in this chapter are valid for arbitrary stationary random field with finite second moment if independence is replaced by orthogonality. In the next chapter we define the multiple Wiener–Itô integrals with respect to this random measure. In the definition of multiple stochastic integrals the Gaussian property will be heavily exploited. First we recall two results about the spectral representation of the covariance function.

Given a stationary Gaussian field X_n, $n \in \mathbb{Z}_\nu$, or $X(\varphi)$, $\varphi \in \mathcal{S}$, we shall assume throughout the paper that $EX_n = 0$, $EX_n^2 = 1$ in the discrete and $EX(\varphi) = 0$ in the generalized field case.

Theorem 3A (Bochner). *Let X_n, $n \in \mathbb{Z}_\nu$, be a discrete (Gaussian) stationary random field. There exists a unique probability measure G on $[-\pi, \pi)^\nu$ such that the correlation function $r(n) = EX_0X_n = EX_kX_{k+n}$, $n \in \mathbb{Z}_\nu$, $k \in \mathbb{Z}_\nu$, can be written in the form*

$$r(n) = \int e^{i(n,x)} G(dx), \tag{3.1}$$

where $(\cdot,\cdot)$ denotes scalar product. Further $G(A) = G(-A)$ for all $A \in [-\pi, \pi)^\nu$.

We can identify $[-\pi, \pi)^\nu$ with the torus $R^\nu / 2\pi\mathbb{Z}_\nu$. Thus e.g. $-(-\pi, \dots, -\pi) = (-\pi, \dots, -\pi)$.

Theorem 3B (Bochner–Schwartz). *Let $X(\varphi)$, $\varphi \in \mathcal{S}$, be a generalized Gaussian stationary random field over $\mathcal{S} = \mathcal{S}_\nu$. There exists a unique σ-finite measure G on R^ν such that*

P. Major, *Multiple Wiener-Itô Integrals*, Lecture Notes in Mathematics 849, DOI 10.1007/978-3-319-02642-8_3,

$$EX(\varphi)X(\psi) = \int \tilde{\varphi}(x)\bar{\tilde{\psi}}(x)G(dx) \qquad \textit{for all } \varphi, \psi \in \mathcal{S}, \tag{3.2}$$

where $\tilde{}$ *denotes Fourier transform and* $\bar{}$ *complex conjugate. The measure G has the properties* $G(A) = G(-A)$ *for all* $A \in \mathcal{B}^\nu$, *and*

$$\int (1+|x|)^{-r} G(dx) < \infty \quad \textit{with an appropriate } r > 0. \tag{3.3}$$

Remark. The above formulated results are actually not the Bochner and Bochner–Schwartz theorem in their original form, they are their consequences. In an Adjustment to Chap. 3 I formulate the classical form of these theorems, and explain how the above results follow from them.

The measure G appearing in Theorems 3A and 3B is called the spectral measure of the stationary field. A measure G with the same properties as the measure G in Theorems 3A or 3B will also be called a spectral measure. This terminology is justified, since there exists a stationary random field with spectral measure G for all such G.

Let us now consider a stationary Gaussian random field (discrete or generalized one) with spectral measure G. We shall denote the space $L_2([-\pi,\pi)^\nu, \mathcal{B}^\nu, G)$ or $L_2(R^\nu, \mathcal{B}^\nu, G)$ simply by L_G^2. Let $\mathcal{H}_1$ denote the real Hilbert space defined by means of the stationary random field, as it was done in Chap. 2. Let $\mathcal{H}_1^c$ denote its complexification, i.e. the elements of $\mathcal{H}_1^c$ are of the form $X + iY$, $X, Y \in \mathcal{H}_1$, and the scalar product is defined as $(X_1 + iY_1, X_2 + iY_2) = EX_1X_2 + EY_1Y_2 + i(EY_1X_2 - EX_1Y_2)$. We are going to construct a unitary transformation I from L_G^2 to $\mathcal{H}_1^c$. We shall define the random spectral measure via this transformation.

Let $\mathcal{S}^c$ denote the Schwartz space of rapidly decreasing, smooth, complex valued functions with the usual topology of the Schwartz space. (The elements of $\mathcal{S}^c$ are of the form $\varphi + i\psi$, $\varphi, \psi \in \mathcal{S}$.) We make the following observation. The finite linear combinations $\sum c_n e^{i(n,x)}$ are dense in L_G^2 in the discrete field, and the functions $\varphi \in \mathcal{S}^c$ are dense in L_G^2 in the generalized field case. In the discrete field case this follows from the Weierstrass approximation theorem, which states that all continuous functions on $[-\pi,\pi)^\nu$ can be approximated arbitrary well in the supremum norm by trigonometrical polynomials. In the generalized field case let us first observe that the continuous functions with compact support are dense in L_G^2. We claim that also the functions of the space $\mathcal{D}$ are dense in L_G^2, where $\mathcal{D}$ denotes the class of (complex valued) infinitely many times differentiable functions with compact support. Indeed, if $\varphi \in \mathcal{D}$ is real valued, $\varphi(x) \ge 0$ for all $x \in R^\nu$, $\int \varphi(x)\,dx = 1$, we define $\varphi_t(x) = t^\nu \varphi\left(\frac{x}{t}\right)$, and f is a continuous function with compact support, then $f * \varphi_t \to f$ uniformly as $t \to \infty$. Here $*$ denotes convolution. On the other hand, $f * \varphi_t \in \mathcal{D}$ for all $t > 0$. Hence $\mathcal{D} \subset \mathcal{S}^c$ is dense in L_G^2.

Finally we recall the following result from the theory of distributions. The mapping $\varphi \to \tilde{\varphi}$ is an invertible, bicontinuous transformation from $\mathcal{S}^c$ into $\mathcal{S}^c$. In particular, the set of functions $\tilde{\varphi}$, $\varphi \in \mathcal{S}$, is also dense in L^2_G.

Now we define the mapping

$$I\left(\sum c_n e^{i(n,x)}\right) = \sum c_n X_n \tag{3.4}$$

in the discrete field case, where the sum is finite, and

$$I(\widetilde{\varphi + i\psi}) = X(\varphi) + iX(\psi), \quad \varphi, \psi \in \mathcal{S} \tag{3.5}$$

in the generalized field case.

Obviously,

$$\begin{aligned}\left\|\sum c_n e^{i(n,x)}\right\|^2_{L^2_G} &= \sum\sum c_n \bar{c}_m \int e^{i(n-m),x} G(dx) \\ &= \sum\sum c_n \bar{c}_m E X_n X_m = E\left|\sum c_n X_n\right|^2,\end{aligned}$$

and

$$\begin{aligned}\|\widetilde{\varphi + i\psi}\|^2_{L^2_G} &= \int [\tilde{\varphi}(x)\bar{\tilde{\varphi}}(x) - i\tilde{\varphi}(x)\bar{\tilde{\psi}}(x) + i\tilde{\psi}(x)\bar{\tilde{\varphi}}(x) + \tilde{\psi}(x)\bar{\tilde{\psi}}(x)] G(dx) \\ &= EX(\varphi)^2 - iEX(\varphi)X(\psi) + iEX(\psi)X(\varphi) + EX(\psi)^2 \\ &= E\left(|X(\varphi) + iX(\psi)|\right)^2.\end{aligned}$$

This means that the mapping I from a linear subspace of L^2_G to $\mathcal{H}_1^c$ is norm preserving. Besides, the subspace where I was defined is dense in L^2_G, since the space of continuous functions is dense in L^2_G if G is a finite measure on the torus $R^\nu/2\pi\mathbb{Z}_\nu$, and the space of continuous functions with a compact support is dense in $L^2_G(R^\nu)$ if the measure G satisfies relation (3.3). Hence the mapping I can be uniquely extended to a norm preserving transformation from L^2_G to $\mathcal{H}_1^c$. Since the random variables X_n or $X(\varphi)$ are obtained as the image of some element from L^2_G under this transformation, I is a unitary transformation from L^2_G to $\mathcal{H}_1^c$. A unitary transformation preserves not only the norm, but also the scalar product. Hence $\int f(x)\bar{g}(x)G(dx) = EI(f)\overline{I(g)}$ for all $f, g \in L^2_G$.

Now we define the random spectral measure $Z_G(A)$ for all $A \in \mathcal{B}^\nu$ such that $G(A) < \infty$ by the formula

$$Z_G(A) = I(\chi_A),$$

where χ_A denotes the indicator function of the set A. It is clear that

(i) The random variables $Z_G(A)$ are complex valued, jointly Gaussian random variables. (The random variables $\operatorname{Re} Z_G(A)$ and $\operatorname{Im} Z_G(A)$ with possibly different sets A are jointly Gaussian.)
(ii) $EZ_G(A) = 0$,
(iii) $EZ_G(A)\overline{Z_G(B)} = G(A \cap B)$,
(iv) $\sum\limits_{j=1}^{n} Z_G(A_j) = Z_G\left(\bigcup\limits_{j=1}^{n} A_j\right)$ if $A_1, \dots, A_n$ are disjoint sets.

Also the following relation holds.

(v) $Z_G(A) = \overline{Z_G(-A)}$.

This follows from the relation

(v′) $I(f) = \overline{I(f_-)}$ for all $f \in L_G^2$, where $f_-(x) = \overline{f(-x)}$.

Relation (v′) can be simply checked if f is a finite trigonometrical polynomial in the discrete field case, or if $f = \tilde{\varphi}$, $\varphi \in \mathcal{S}^c$, in the generalized field case. (In the case $f = \tilde{\varphi}$, $\varphi \in \mathcal{S}^c$, the following argument works. Put $f(x) = \tilde{\varphi}_1(x) + i\tilde{\varphi}_2(x)$ with $\varphi_1, \varphi_2 \in \mathcal{S}$. Then $I(f) = X(\varphi_1) + iX(\varphi_2)$, and $f_-(x) = \bar{\tilde{\varphi}}_1(-x) - i\bar{\tilde{\varphi}}_2(-x) = \tilde{\varphi}_1(x) + i(\widetilde{-\varphi_2}(x)$, hence $I(f_-) = X(\varphi_1) + iX(-\varphi_2) = X(\varphi_1) - iX(\varphi_2) = \overline{I(f)}$.) Then a simple limiting procedure implies (v′) in the general case. Relation (iii) follows from the identity $EZ_G(A)\overline{Z_G(B)} = EI(\chi_A)\overline{I(\chi_B)} = \int \chi_A(x)\overline{\chi_B(x)}G(dx) = G(A \cap B)$. The remaining properties of $Z_G(\cdot)$ are simple consequences of the definition.

Remark. Property (iv) could have been omitted from the definition of random spectral measures, since it follows from property (iii). To show this it is enough to check that if $A_1, \dots, A_n$ are disjoint sets, and property (iii) holds, then

$$E\left(\sum_{j=1}^{n} Z_G(A_j) - Z_G\left(\bigcup_{j=1}^{n} A_j\right)\right)\overline{\left(\sum_{j=1}^{n} Z_G(A_j) - Z_G\left(\bigcup_{j=1}^{n} A_j\right)\right)} = 0.$$

Now we introduce the following

Definition of Random Spectral Measure. *Let G be a spectral measure. A set of random variables $Z_G(A)$, $G(A) < \infty$, satisfying (i)–(v) is called a (Gaussian) random spectral measure corresponding to the spectral measure G.*

Given a Gaussian random spectral measure Z_G corresponding to a spectral measure G we define the (onefold) stochastic integral $\int f(x) Z_G(dx)$ for an appropriate class of functions f. Let us first consider simple functions of the form $f(x) = \sum c_i \chi_{A_i}(x)$, where the sum is finite, and $G(A_i) < \infty$ for all indices i. In this case we define

$$\int f(x) Z_G(dx) = \sum c_i Z_G(A_i).$$

Then we have

$$E\left|\int f(x)Z_G(dx)\right|^2 = \sum c_i\bar{c}_j G(A_i \cap A_j) = \int |f(x)|^2 G(dx). \tag{3.6}$$

Since the simple functions are dense in L_G^2, relation (3.6) enables us to define $\int f(x)Z_G(dx)$ for all $f \in L_G^2$ via L_2-continuity. It can be seen that this integral satisfies the identity

$$E\int f(x)Z_G(dx)\overline{\int g(x)Z_G(dx)} = \int f(x)\overline{g(x)}G(dx) \tag{3.7}$$

for all pairs of functions $f, g \in L_G^2$. Moreover, similar approximation with simple functions yields that

$$\int f(x)Z_G(dx) = \overline{\int \overline{f(-x)}Z_G(dx)} \tag{3.8}$$

for a function $f \in L_G^2$. Here we exploit the identity $Z_G(A) = \overline{Z_G(-A)}$ formulated in property (v) of the random spectral measure Z_G.

The last two identities together with the relations (3.1) and (3.2) imply that if we define the set of random variables X_n and $X(\varphi)$ by means of the formula

$$X_n = \int e^{i(n,x)} Z_G(dx), \quad n \in \mathbb{Z}_\nu, \tag{3.9}$$

and

$$X(\varphi) = \int \tilde{\varphi}(x) Z_G(dx), \quad \varphi \in \mathcal{S}, \tag{3.10}$$

where we integrate with respect to the random spectral measure Z_G, then we get a Gaussian stationary random discrete and generalized field with spectral measure G, i.e. with correlation function given in formulas (3.1) and (3.2). To check this statement first we have to show that the random variables X_n and $X(\varphi)$ defined in (3.9) and (3.10) are real valued, or equivalently saying the identities $X_n = \overline{X_n}$ and $X(\varphi) = \overline{X(\varphi)}$ hold with probability 1. This follows from relation (3.8) and the identities $e^{i(n,x)} = \overline{e^{(i(n,-x)}}$ and $\tilde{\varphi}(x) = \overline{\tilde{\varphi}(-x)}$ for a (real valued) function $\varphi \in \mathcal{S}$. Then we can calculate the correlation functions $EX_nX_m = EX_n\overline{X}_m$ and $EX(\varphi)X(\psi) = EX(\varphi)\overline{X(\psi)}$ by means of formula (3.7), (3.9) and (3.10).

We also have

$$\int f(x)Z_G(dx) = I(f) \quad \text{for all } f \in L_G^2$$

if we consider the previously defined mapping $I(f)$ with the stationary random fields defined in (3.9) and (3.10). Now we formulate the following

Theorem 3.1. *For a stationary Gaussian random field (a discrete or generalized one) with a spectral measure G there exists a unique Gaussian random spectral measure Z_G corresponding to the spectral measure G on the same probability space as the Gaussian random field such that relation (3.9) or (3.10) holds in the discrete or generalized field case respectively.*

Furthermore

$$\mathcal{B}(Z_G(A),\ G(A) < \infty) = \begin{cases} \mathcal{B}(X_n,\ n \in \mathbb{Z}_\nu) \text{ in the discrete field case,} \\ \mathcal{B}(X(\varphi),\ \varphi \in \mathcal{S}) \text{ in the generalized field case.} \end{cases} \tag{3.11}$$

If a stationary Gaussian random field $X_n, n \in \mathbb{Z}_\nu$, or $X(\varphi), \varphi \in \mathcal{S}$, and a random spectral measure Z_G satisfy relation (3.9) or (3.10), then we say that this random spectral measure is adapted to this Gaussian random field.

Proof of Theorem 3.1. Given a stationary Gaussian random field (discrete or stationary one) with a spectral measure G, we have constructed a random spectral measure Z_G corresponding to the spectral measure G. Moreover, the random integrals given in formulas (3.9) or (3.10) define the original stationary random field. Since all random variables $Z_G(A)$ are measurable with respect to the original random field, relation (3.9) or (3.10) implies (3.11).

To prove the uniqueness, it is enough to observe that because of the linearity and L_2 continuity of stochastic integrals relation (3.9) or (3.10) implies that

$$Z_G(A) = \int \chi_A(x) Z_G(dx) = I(\chi_A)$$

for a Gaussian random spectral measure corresponding to the spectral measure G appearing in Theorem 3.1. □

Finally we list some additional properties of Gaussian random spectral measures.

(vi) The random variables $\operatorname{Re} Z_G(A)$ are independent of the random variables $\operatorname{Im} Z_G(A)$.
(vii) The random variables of the form $Z_G(A \cup (-A))$ are real valued. If the sets $A_1 \cup (-A_1), \dots, A_n \cup (-A_n)$ are disjoint, then the random variables $Z_G(A_1), \dots, Z_G(A_n)$ are independent.
(viii) The relations $\operatorname{Re} Z_G(-A) = \operatorname{Re} Z_G(A)$ and $\operatorname{Im} Z_G(-A) = -\operatorname{Im} Z_G(A)$ hold, and if $A \cap (-A) = \emptyset$, then the (Gaussian) random variables $\operatorname{Re} Z_G(A)$ and $\operatorname{Im} Z_G(A)$ are independent with expectation zero and variance $G(A)/2$.

These properties easily follow from (i)–(v). Since $Z_G(\cdot)$ are complex valued Gaussian random variables, to prove the above formulated independence it is enough to show that the real and imaginary parts are uncorrelated. We show, as an example, the proof of (vi).

$$
\begin{aligned}
E\mathrm{Re}\, Z_G(A)\mathrm{Im}\, Z_G(B) &= \frac{1}{4i}E(Z_G(A)+\overline{Z_G(A)})(Z_G(B)-\overline{Z_G(B)})\\
&= \frac{1}{4i}E(Z_G(A)+Z_G(-A))(\overline{Z_G(-B)}-\overline{Z_G(B)})\\
&= \frac{1}{4i}G(A\cap(-B))-\frac{1}{4i}G(A\cap B)\\
&\quad +\frac{1}{4i}G((-A)\cap(-B))-\frac{1}{4i}G((-A)\cap B)=0
\end{aligned}
$$

for all pairs of sets A and B such that $G(A)<\infty$, $G(B)<\infty$, since $G(D)=G(-D)$ for all $D\in\mathscr{B}^\nu$. The fact that $Z_G(A\cup(-A))$ is real valued random variable, and the relations $\mathrm{Re}\, Z_G(-A)=\mathrm{Re}\, Z_G(A)$, $\mathrm{Im}\, Z_G(-A)=-\mathrm{Im}\, Z_G(A)$ under the conditions of (viii) follow directly from (v). The remaining statements of (vii) and (viii) can be proved similarly to (vi) only the calculations are simpler in this case.

The properties of the random spectral measure Z_G listed above imply in particular that the spectral measure G determines the joint distribution of the corresponding random variables $Z_G(B)$, $B\in\mathscr{B}^\nu$.

3.1 On the Spectral Representation of the Covariance Function of Stationary Random Fields

The results formulated under the name of Bochner and Bochner–Schwartz theorem (I write this, because actually I presented not these theorems but an important consequence of them) have the following content. Given a finite, even measure G on the torus $R^\nu/2\pi\mathbb{Z}_\nu$ one can define a (Gaussian) discrete stationary field with correlation function satisfying (3.1) with this measure G. For an even measure G on R^ν satisfying (3.3) there exists a (Gaussian) generalized stationary field with correlation function defined in formula (3.2) with this measure G. The Bochner and Bochner–Schwartz theorems state that the correlation function of all (Gaussian) discrete stationary fields, respectively of all stationary generalized fields can be represented in such a way. Let us explain this in more detail.

First I formulate the following

Proposition 3C. *Let G be a finite measure on the torus $R^\nu/2\pi\mathbb{Z}_\nu$ such that $G(A)=G(-A)$ for all measurable sets A. Then there exists a Gaussian discrete stationary random field X_n, $n\in\mathbb{Z}_\nu$, with expectation zero such that its correlation function $r(n)=EX_kX_{k+n}$, $n,k\in\mathbb{Z}_\nu$, is given by formula (3.1) with this measure G.*

Let G be a measure on R^ν satisfying (3.3) and such that $G(A)=G(-A)$ for all measurable sets A. Then there exists a Gaussian stationary generalized random field

$X(\varphi)$, $\varphi \in \mathcal{S}$, with expectation $EX(\varphi) = 0$ for all $\varphi \in \mathcal{S}$ such that its covariance function $EX(\varphi)X(\psi)$, $\varphi, \psi \in \mathcal{S}$, satisfies formula (3.2) with this measure G.

Moreover, the correlation function $r(n)$ or $EX(\varphi)X(\psi)$, $\varphi, \psi \in \mathcal{S}$, determines the measure G uniquely.

Proof of Proposition 3C. By Kolmogorov's theorem about the existence of random processes with consistent finite dimensional distributions it is enough to prove the following statement to show the existence of the Gaussian discrete stationary field with the demanded properties. For any points $n_1, \dots, n_p \in \mathbb{Z}_\nu$ there exists a Gaussian random vector $(X_{n_1}, \dots, X_{n_p})$ with expectation zero and covariance matrix $EX_{n_j}X_{n_k} = r(n_j - n_k)$. (Observe that the function $r(n)$ is real valued, $r(n) = r(-n)$, because of the evenness of the spectral measure G.) Hence it is enough to check that the corresponding matrix is positive definite, i.e. $\sum\limits_{j,k} c_j c_k r(n_j - n_k) \ge 0$ for all real vectors $(c_1, \dots, c_p)$. This relation holds, because $\sum\limits_{j,k} c_j c_k r(n_j - n_k) = \int |\sum\limits_j c_j e^{i(n_j,x)}|^2\, G(dx) \ge 0$ by formula (3.1).

It can be proved similarly that in the generalized field case there exists a Gaussian random field with expectation zero whose covariance function satisfies formula (3.2). (Let us observe that the relation $G(A) = G(-A)$ implies that $EX(\varphi)X(\psi)$ is a real number for all $\varphi, \psi \in \mathcal{S}$, since $EX(\varphi)X(\psi) = \overline{EX(\varphi)X(\psi)}$ in this case. In the proof of this identity we exploit that $\bar{\tilde{f}}(x) = \tilde{f}(-x)$ for a real valued function f.) We also have to show that a random field with such a distribution is a generalized field, i.e. it satisfies properties (a) and (b) given in the definition of generalized fields.

It is not difficult to show that if $\varphi_n \to \varphi$ in the topology of the space $\mathcal{S}$, then $E[X(\varphi_n) - X(\varphi)]^2 = \int |\tilde{\varphi}_n(x) - \tilde{\varphi}(x)|^2 G(dx) \to 0$ as $n \to \infty$, hence property (b) holds. (Here we exploit that the transformation $\varphi \to \tilde{\varphi}$ is bicontinuous in the space $\mathcal{S}$.) Property (a) also holds, because, as it is not difficult to check with the help of formula (3.2),

$$\begin{aligned} &E[a_1X(\varphi_1) + a_2X(\varphi_2) - X(\varphi(a_1\varphi_1 + a_2\varphi_2))]^2 \\ &\quad = \int \left| a_1\tilde{\varphi}_1(x) + a_2\tilde{\varphi}_2(x) - (\widetilde{a_1\varphi_1 + a_2\varphi_2})(x)\right|^2 G(dx) = 0. \end{aligned}$$

It is clear that the Gaussian random field constructed in such a way is stationary.

Finally, as we have seen in our considerations in the main text, the correlation function determines the integral $\int f(x)\, G(dx)$ for all continuous functions f with a bounded support, hence it also determines the measure G. □

The Bochner and Bochner–Schwartz theorems enable us to show that the correlation function of all stationary (Gaussian) random fields (discrete or generalized one) can be represented in the above way with an appropriate spectral measure G. To see this let us formulate these results in their original form.

To formulate Bochner's theorem first we introduce the following notion.

Definition of Positive Definite Functions. *Let $f(x)$ be a (complex valued) function on $\mathbb{Z}_\nu$ (or on R^ν). We say that $f(\cdot)$ is a positive definite function if for all parameters p, complex numbers $c_1,\dots,c_p$ and points $x_1,\dots,x_p$ in $\mathbb{Z}_\nu$ (or in R^ν) the inequality*

$$\sum_{j=1}^{p}\sum_{k=1}^{p} c_j\bar{c}_k f(x_j-x_k)\ge 0$$

holds.

A simple example for positive definite functions is the function $f(x)=e^{i(t,x)}$, where $t\in\mathbb{Z}_\nu$ in the discrete, and $t\in R^\nu$ in the continuous case. Bochner's theorem provides a complete description of positive definite functions.

Bochner's Theorem. (Its Original Form.) *A complex valued function $f(x)$ defined on $\mathbb{Z}_\nu$ is positive definite if and only if it can be written in the form $f(x)=\int e^{i(t,x)}G(dx)$ for all $x\in\mathbb{Z}_\nu$ with a finite measure G on the torus $R^\nu/2\pi\mathbb{Z}_\nu$. The measure G is uniquely determined.*

A complex valued function $f(x)$ defined on R^ν is positive definite and continuous at the origin if and only if it can be written in the form $f(x)=\int e^{i(t,x)}G(dx)$ for all $x\in R^\nu$ with a finite measure G on R^ν. The measure G is uniquely determined.

It is not difficult to see that the covariance function $r(n)=EX_kX_{k+n}$, $(EX_n=0)$, $k,n\in\mathbb{Z}_\nu$, of a stationary (Gaussian) random field X_n is a positive definite function, since $\sum\limits_{j,k} c_j\bar{c}_k r(n_j-n_k)=E|\sum\limits_j c_jX_{n_j}|^2>0$ for any vector $(c_1,\dots,c_p)$. Hence Bochner's theorem can be applied for it. Besides, the relation $r(n)=r(-n)$ together with the uniqueness of the measure G appearing in Bochner's theorem imply that the identity $G(A)=G(-A)$ holds for all measurable sets G. This implies the result formulated in the main text under the name Bochner's theorem.

The Bochner–Schwartz theorem yields an analogous representation of positive definite generalized functions in $\mathcal{S}'$ as the Fourier transforms of positive generalized functions in $\mathcal{S}'$. It also states a similar result about generalized functions in the space $\mathcal{D}'$. To formulate it we have to introduce some definitions. First we have to clarify what a positive generalized function is. We introduce this notion both in the space $\mathcal{S}'$ and $\mathcal{D}'$, and then we characterize them in a Theorem.

Definition of Positive Generalized Functions. *A continuous linear functional $F\in\mathcal{S}'$ (or $F\in\mathcal{D}'$) is called a positive generalized function if for all such $\varphi\in\mathcal{S}$ (or $\varphi\in\mathcal{D}$) test functions for which $\varphi(x)\ge 0$ for all $x\in R^\nu$ $(F,\varphi)\ge 0$.*

Theorem About the Representation of Positive Generalized Functions. *All positive generalized functions $F\in\mathcal{S}'$ can be given in the form $(F,\varphi)=\int\varphi(x)\mu(dx)$, where μ is a polynomially increasing measure on R^ν, i.e. it satisfies the relation $\int(1+|x|^2)^{-p}\mu(dx)<\infty$ with some $p>0$. Similarly, all positive generalized functions in $\mathcal{D}'$ can be given in the form $(F,\varphi)=\int\varphi(x)\mu(dx)$*

with such a measure μ on R^ν which is finite in all bounded regions. The positive generalized function F uniquely determines the measure μ in both cases.

We also introduce a rather technical notion and formulate a result about it. Let us remark that if $\varphi \in \mathscr{S}^c$ and $\psi \in \mathscr{S}^c$, then also their product $\varphi\psi \in \mathscr{S}^c$. In particular, $\varphi\bar{\varphi} = |\varphi|^2 \in \mathscr{S}$ if $\varphi \in \mathscr{S}^c$. The analogous result also holds in the space $\mathscr{D}$.

Definition of Multiplicatively Positive Generalized Functions. *A generalized function $\mathscr{F} \in \mathscr{S}'$ (or $F \in \mathscr{D}'$) is multiplicatively positive if $(F, \varphi\bar{\varphi}) = (F, |\varphi|^2) \geq 0$ for all $\varphi \in \mathscr{S}^c$ (or in $\varphi \in \mathscr{D}$).*

Theorem About the Characterization of Multiplicatively Positive Generalized Functions. *A generalized function $F \in \mathscr{S}'$ (or $F \in \mathscr{D}'$) is multiplicatively positive if and only if it is positive.*

Now I introduce the definition of positive definite generalized functions.

Definition of Positive Definite Generalized Functions. A generalized function $F \in \mathscr{S}'$ (or $F \in \mathscr{D}'$) is positive definite if $(F, \varphi * \varphi^*) \geq 0$ for all $\varphi \in \mathscr{S}^c$ (of $\varphi \in \mathscr{D}$), where $\varphi^*(x) = \overline{\varphi(-x)}$, and $*$ denotes convolution, i.e. $\varphi * \varphi^*(x) = \int \varphi(t)\overline{\varphi(t-x)}\,dt$.

We refer to [16] for an explanation why this definition of positive definite generalized functions is natural. Let us remark that if $\varphi, \psi \in \mathscr{S}^c$, then $\varphi * \psi \in \mathscr{S}^c$, and the analogous result holds in $\mathscr{D}$. The original version of the Bochner–Schwartz theorem has the following form.

Bochner–Schwartz Theorem. (Its Original Form.) *Let F be a positive definite generalized function in the space $\mathscr{S}'$ (or $\mathscr{D}'$). Then it is the Fourier transform of a polynomially increasing measure μ on R^ν, i.e. the identity $(F, \varphi) = \int \tilde{\varphi}(x)\,\mu(dx)$ holds for all $\varphi \in \mathscr{S}^c$ (or $\varphi \in \mathscr{D}$) with a measure μ that satisfies the relation $\int (1 + |x|^2)^{-p}\mu(dx) < \infty$ with an appropriate $p > 0$. The generalized function F uniquely determines the measure μ. On the other hand, if μ is a polynomially increasing measure on R^ν, then the formula $(F, \varphi) = \int \tilde{\varphi}(x)\mu(dx)$ with $\varphi \in \mathscr{S}^c$ (or $\varphi \in \mathscr{D}$) defines a positive definite generalized function F in the space $\mathscr{S}'$ (or $\mathscr{D}'$).*

Remark. It is a remarkable and surprising fact that the class of positive definite generalized functions are represented by the same class of measures μ in the spaces $\mathscr{S}'$ and $\mathscr{D}'$. (In the representation of positive generalized functions the class of measures μ considered in the case of $\mathscr{D}'$ is much larger, than in the case of $\mathscr{S}'$.) Let us remark that in the representation of the positive definite generalized functions in $\mathscr{D}'$ the function $\tilde{\varphi}$ we integrate is not in the class $\mathscr{D}$, but in the space $\mathscr{Z}$ consisting of the Fourier transforms of the functions in $\mathscr{D}$.

It is relatively simple to prove the representation of positive definite generalized functions given in the Bochner–Schwartz theorem for the class $\mathscr{S}'$. Some calculation shows that if F is a positive definite generalized function, then its Fourier transform is a multiplicatively positive generalized function. Indeed, since the Fourier transform of the convolution $\varphi * \psi(x)$ equals $\tilde{\varphi}(t)\tilde{\psi}(t)$, and the Fourier transform of $\varphi^*(x) = \overline{\varphi(-x)}$ equals $\overline{\tilde{\varphi}(t)}$, the Fourier transform of $\varphi * \varphi^*(x)$ equals $\tilde{\varphi}(t)\bar{\tilde{\varphi}}(t)$. Hence the positive definitiveness property of the generalized function F and the definition of the Fourier transform of generalized functions imply that $(\tilde{F}, \tilde{\varphi}\bar{\tilde{\varphi}}) = (2\pi)^\nu (F, \varphi * \varphi^*) \geq 0$ for all $\varphi \in \mathcal{S}^c$. Since every function of $\mathcal{S}^c$ is the Fourier transform $\tilde{\varphi}$ of some function $\varphi \in \mathcal{S}^c$ this implies that $\tilde{F}$ is a multiplicatively positive and as a consequence a positive generalized function in $\mathcal{S}'$. Such generalized functions have a good representation with the help of a polynomially increasing positive measure μ. Since $(F, \varphi) = (2\pi)^{-\nu}(\tilde{F}, \tilde{\varphi})$ it is not difficult to prove the Bochner–Schwartz theorem for the space $\mathcal{S}'$ with the help of this fact. The proof is much harder if the space $\mathcal{D}'$ is considered, but we do not need that result.

The Bochner–Schwartz theorem in itself is not sufficient to describe the correlation function of a generalized random field. We still need another important result of Laurent Schwartz which gives useful information about the behaviour of (Hermitian) bilinear functionals in $\mathcal{S}^c$ and some additional information about the behaviour of translation invariant (Hermitian) bilinear functionals in this space. To formulate these results first we introduce the following definition.

Definition of Hermitian Bilinear and Translation Invariant Hermitian Bilinear Functionals in the Space $\mathcal{S}^c$. *A function $B(\varphi, \psi)$, $\varphi, \psi \in \mathcal{S}^c$, is a Hermitian bilinear functional in the space $\mathcal{S}^c$ if for all fixed $\psi \in \mathcal{S}^c$ $B(\varphi, \psi)$ is a continuous linear functional of the variable φ in the topology of $\mathcal{S}^c$, and for all fixed $\varphi \in \mathcal{S}^c$ $B(\varphi, \psi)$ is a continuous linear functional of the variable ψ in the topology of $\mathcal{S}^c$.*

A Hermitian bilinear functional $B(\varphi, \psi)$ in $\mathcal{S}^c$ is translation invariant if it does not change by a simultaneous shift of its variables φ and ψ, i.e. if $B(\varphi(x), \psi(x)) = B(\varphi(x-h), \psi(x-h))$ for all $h \in R^\nu$.

Definition of Positive Definite Hermitian Bilinear Functionals. *We say that a Hermitian bilinear functional $B(\varphi, \psi)$ in $\mathcal{S}^c$ is positive definite if $B(\varphi, \varphi) \geq 0$ for all $\varphi \in \mathcal{S}^c$.*

The next result characterizes the Hermitian bilinear and translation invariant Hermitian bilinear functionals in $\mathcal{S}^c$.

Theorem 3D. *All Hermitian bilinear functionals $B(\varphi, \psi)$ in $\mathcal{S}^c$ can be given in the form $B(\varphi, \psi) = (F_1, \varphi(x)\overline{\psi(y)})$, $\varphi, \psi \in \mathcal{S}^c$, where F_1 is a continuous linear functional on $\mathcal{S}^c \times \mathcal{S}^c$, i.e. it is a generalized function in $\mathcal{S}_{2\nu}'$.*

A translation invariant Hermitian bilinear functional in $\mathcal{S}^c$ can be given in the form $\mathcal{B}(\varphi, \psi) = (F, \varphi * \psi^*)$, $\varphi, \psi \in \mathcal{S}^c$, where $F \in \mathcal{S}$, $\psi^*(x) = \overline{\psi}(-x)$, and $*$ denotes convolution.

The Hermitian bilinear form $B(\varphi, \psi)$ determines the generalized functions F_1 uniquely, and if it is translation invariant, then the same can be told about the generalized function F. Besides, for all functionals $F_1 \in \mathcal{S}_{2\nu}'$ and $F \in \mathcal{S}'$ the above formulas define a Hermitian bilinear functional and a translation invariant Hermitian bilinear functional in $\mathcal{S}_\nu^c$ respectively.

Let us consider a Gaussian generalized random field $X(\varphi)$, $\varphi \in \mathcal{S}$, with expectation zero together with its correlation function $B(\varphi, \psi) = EX\varphi)X(\psi)$, $\varphi, \psi \in \mathcal{S}$. More precisely, let us consider the complexification $X(\varphi_1 + i\varphi_2) = X(\varphi_1) + iX(\varphi_2)$ of this random field and its correlation function $B(\varphi, \psi) = EX(\varphi)\overline{X(\psi)}$, $\varphi, \psi \in \mathcal{S}^c$. This correlation function $B(\varphi, \psi)$ is a translation invariant Hermitian bilinear functional in $\mathcal{S}^c$, hence it can be written in the form $B(\varphi, \psi) = (F, \varphi * \psi^*)$ with an appropriate $F \in \mathcal{S}'$. Moreover, $B(\varphi, \varphi) \geq 0$ for all $\varphi \in \mathcal{S}^c$, and this means that the generalized function $F \in \mathcal{S}'$ corresponding to $B(\varphi, \psi)$ is positive definite. Hence the Bochner–Schwartz theorem can be applied for it, and it yields that

$$EX(\varphi)X(\psi) = \int \widetilde{\varphi * \psi^*}(x)\, G(dx) = \int \tilde{\varphi}(x)\bar{\tilde{\psi}}(x)\, G(dx) \quad \text{for all } \varphi,\ \psi \in \mathcal{S}^c$$

with a uniquely determined, polynomially increasing measure G on R^ν. Now we complete the proof of Theorem 3B with the help of these results.

Proof of Theorem 3B. We have already proved relations (3.2) and (3.3) with the help of some results about generalized functions. To complete the proof of Theorem 3B we still have to show that G is an even measure. In the proof of this statement we exploit that for a real valued function $\varphi \in \mathcal{S}$ the random variable $X(\varphi)$ is also real valued. Hence if $\varphi, \psi \in \mathcal{S}$, then $EX(\varphi)X(\psi) = \overline{EX(\varphi)X(\psi)}$. Besides, $\tilde{\varphi}(-x) = \bar{\tilde{\varphi}}(x)$ and $\tilde{\psi}(-x) = \bar{\tilde{\psi}}(x)$ in this case. Hence

$$\begin{aligned}\int \tilde{\varphi}(x)\bar{\tilde{\psi}}(x)\, G(dx) &= \int \bar{\tilde{\varphi}}(x)\tilde{\psi}(x)\, G(dx)\\ &= \int \tilde{\varphi}(-x)\bar{\tilde{\psi}}(-x)\, G(dx) = \int \tilde{\varphi}(x)\bar{\tilde{\psi}}(x)\, G^-(dx)\end{aligned}$$

for all $\varphi, \psi \in \mathcal{S}$, where $G^-(A) = G(-A)$ for all $A \in \mathcal{B}^\nu$. This relation implies that the measures G and G^- agree. The proof of Theorem 3B is completed. □

Chapter 4
Multiple Wiener–Itô Integrals

In this chapter we define the so-called multiple Wiener–Itô integrals, and we prove their most important properties with the help of Itô's formula, whose proof is postponed to the next chapter. More precisely, we discuss in this chapter a modified version of the Wiener–Itô integrals with respect to a random spectral measure rather than with respect to a random measure with independent increments. This modification makes it necessary to slightly change the definition of the integral. This modified Wiener–Itô integral seems to be a more useful tool than the original one or the Wick polynomials in the study of the problems in this work, because it enables us to describe the action of shift transformations.

Let G be the spectral measure of a stationary Gaussian field (discrete or generalized one). We define the following *real* Hilbert spaces $\bar{\mathscr{H}}_G^n$ and $\mathscr{H}_G^n$, $n = 1, 2, \dots$. We have $f_n \in \bar{\mathscr{H}}_G^n$ if and only if $f_n = f_n(x_1, \dots, x_n)$, $x_j \in R^\nu$, $j = 1, 2, \dots, n$, is a complex valued function of n variables, and

(a) $f_n(-x_1, \dots, -x_n) = \overline{f_n(x_1, \dots, x_n)}$,
(b) $\|f_n\|^2 = \int |f_n(x_1, \dots, x_n)|^2 G(dx_1) \dots G(dx_n) < \infty$.

Relation (b) also defines the norm in $\bar{\mathscr{H}}_G^n$. The subspace $\mathscr{H}_G^n \subset \bar{\mathscr{H}}_G^n$ contains those functions $f_n \in \bar{\mathscr{H}}_G^n$ which are invariant under permutations of their arguments, i.e.

(c) $f_n(x_{\pi(1)}, \dots, x_{\pi(n)})) = f_n(x_1, \dots, x_n)$ for all $\pi \in \Pi_n$, where Π_n denotes the group of all permutations of the set $\{1, 2, \dots, n\}$.

The norm in $\mathscr{H}_G^n$ is defined in the same way as in $\bar{\mathscr{H}}_G^n$. Moreover, the scalar product is also similarly defined, namely if $f, g \in \bar{\mathscr{H}}_G^n$, then

$$
\begin{aligned}
(f, g) &= \int f(x_1, \dots, x_n) \overline{g(x_1, \dots, x_n)} G(dx_1) \dots G(dx_n) \\
&= \int f(x_1, \dots, x_n) g(-x_1, \dots, -x_n) G(dx_1) \dots G(dx_n).
\end{aligned}
$$

P. Major, *Multiple Wiener-Itô Integrals*, Lecture Notes in Mathematics 849, DOI 10.1007/978-3-319-02642-8_4,

Because of the symmetry $G(A) = G(-A)$ of the spectral measure $(f, g) = \overline{(f, g)}$, i.e. the scalar product (f, g) is a real number for all $f, g \in \bar{\mathcal{H}}_G^n$. This means that $\bar{\mathcal{H}}_G^n$ is a real Hilbert space. We also define $\mathcal{H}_G^0 = \bar{\mathcal{H}}_G^0$ as the space of real constants with the norm $\|c\| = |c|$. We remark that $\bar{\mathcal{H}}_G^n$ is actually the n-fold direct product of $\mathcal{H}_G^1$, while $\mathcal{H}_G^n$ is the n-fold symmetrical direct product of $\mathcal{H}_G^1$. Condition (a) means heuristically that f_n is the Fourier transform of a real valued function.

Finally we define the so-called Fock space $\operatorname{Exp} \mathcal{H}_G$ whose elements are sequences of functions $f = (f_0, f_1, \dots)$, $f_n \in \mathcal{H}_G^n$ for all $n = 0, 1, 2, \dots,$ such that

$$\|f\|^2 = \sum_{n=0}^{\infty} \frac{1}{n!} \|f_n\|^2 < \infty.$$

Given a function $f \in \bar{\mathcal{H}}_G^n$ we define $\operatorname{Sym} f$ as

$$\operatorname{Sym} f(x_1, \dots, x_n) = \frac{1}{n!} \sum_{\pi \in \Pi_n} f(x_{\pi(1)}, \dots, x_{\pi(n)}).$$

Clearly, $\operatorname{Sym} f \in \mathcal{H}_G^n$, and

$$\|\operatorname{Sym} f\| \le \|f\|. \tag{4.1}$$

Let Z_G be a Gaussian random spectral measure corresponding to the spectral measure G on a probability space $(\Omega, \mathcal{A}, P)$. We shall define the n-fold Wiener–Itô integrals

$$I_G(f_n) = \frac{1}{n!} \int f_n(x_1, \dots, x_n) Z_G(dx_1) \dots Z_G(dx_n) \quad \text{for } f_n \in \bar{\mathcal{H}}_G^n$$

and

$$I_G(f) = \sum_{n=0}^{\infty} I_G(f_n) \quad \text{for } f = (f_0, f_1, \dots) \in \operatorname{Exp} \mathcal{H}_G.$$

We shall see that $I_G(f_n) = I_G(\operatorname{Sym} f_n)$ for all $f_n \in \bar{\mathcal{H}}_G^n$. Therefore, it would have been sufficient to define the Wiener–Itô integral only for functions in $\mathcal{H}_G^n$. Nevertheless, some arguments become simpler if we work in $\bar{\mathcal{H}}_G^n$. In the definition of Wiener–Itô integrals first we restrict ourselves to the case when the spectral measure is non-atomic, i.e. $G(\{x\}) = 0$ for all $x \in R^\nu$. This condition is satisfied in all interesting cases. However, we shall later show how one can get rid of this restriction.

First we introduce the notion of regular systems for some collections of subsets of R^ν, define a subclass $\hat{\bar{\mathcal{H}}}_G^n \subset \bar{\mathcal{H}}_G^n$ of simple functions with their help, and define the Wiener–Itô integrals for the functions of this subclass.

Definition of Regular Systems and the Class of Simple Functions. *Let*

$$\mathcal{D} = \{\Delta_j, \ j = \pm 1, \pm 2, \dots, \pm N\}$$

be a finite collection of bounded, measurable sets in R^ν indexed by the integers $\pm 1, \dots, \pm N$. We say that $\mathcal{D}$ is a regular system if $\Delta_j = -\Delta_{-j}$, and $\Delta_j \cap \Delta_l = \emptyset$ if $j \neq l$ for all $j, l = \pm 1, \pm 2, \dots, \pm N$. A function $f \in \bar{\mathcal{H}}_G^n$ is adapted to this system $\mathcal{D}$ if $f(x_1, \dots, x_n)$ is constant on the sets $\Delta_{j_1} \times \Delta_{j_2} \times \dots \times \Delta_{j_n}$, $j_l = \pm 1, \dots, \pm N$, $l = 1, 2, \dots, n$, it vanishes outside these sets and also on those sets of the form $\Delta_{j_1} \times \Delta_{j_2} \times \dots \times \Delta_{j_n}$, for which $j_l = \pm j_{l'}$ for some $l \neq l'$.

A function $f \in \bar{\mathcal{H}}_G^n$ is in the class $\hat{\bar{\mathcal{H}}}_G^n$ of simple functions, and a (symmetric) function $f \in \mathcal{H}_G^n$ is in the class $\hat{\mathcal{H}}_G^n$ of simple symmetric functions if it is adapted to some regular system $\mathcal{D} = \{\Delta_j, \ j = \pm 1, \dots, \pm N\}$.

Definition of Wiener–Itô Integral of Simple Functions. *Let a simple function $f \in \hat{\bar{\mathcal{H}}}_G^n$ be adapted to some regular systems $\mathcal{D} = \{\Delta_j, \ j \pm 1, \dots, \pm N\}$. Its Wiener–Itô integral with respect to the random spectral measure Z_G is defined as*

$$\int f(x_1, \dots, x_n) Z_G(dx_1) \dots Z_G(dx_n) \tag{4.2}$$

$$= n! I_G(f) = \sum_{\substack{j_l = \pm 1, \dots, \pm N \\ l = 1, 2, \dots, n}} f(x_{j_1}, \dots, x_{j_n}) Z_G(\Delta_{j_1}) \cdots Z_G(\Delta_{j_n}),$$

where $x_{j_l} \in \Delta_{j_l}$, $j_l = \pm 1, \dots, \pm N$, $l = 1, \dots, n$.

We remark that although the regular system $\mathcal{D}$ to which f is adapted, is not uniquely determined (the elements of $\mathcal{D}$ can be divided to smaller sets), the integral defined in (4.2) is meaningful, i.e. it does not depend on the choice of $\mathcal{D}$. This can be seen by observing that a refinement of a regular system $\mathcal{D}$ to which the function f is adapted yields the same value for the sum defining $n! I_G(f)$ in formula (4.2) as the original one. This follows from the additivity of the random spectral measure Z_G formulated in its property (iv), since this implies that each term $f(x_{j_1}, \dots, x_{j_n}) Z_G(\Delta_{j_1}) \cdots Z_G(\Delta_{j_n})$ in the sum at the right-hand side of formula (4.2) corresponding to the original regular system equals the sum of all such terms $f(x_{j_1}, \dots, x_{j_n}) Z_G(\Delta'_{j'_1}) \cdots Z_G(\Delta'_{j'_n})$ in the sum corresponding to the refined partition for which $\Delta'_{j'_1} \times \dots \times \Delta'_{j'_n} \subset \Delta_{j_1} \times \dots \times \Delta_{j_n}$.

By property (vii) of the random spectral measures all products

$$Z_G(\Delta_{j_1}) \cdots Z_G(\Delta_{j_n})$$

with non-zero coefficient in (4.2) are products of independent random variables. We had this property in mind when requiring the condition that the function f vanishes on a product $\Delta_{j_1} \times \cdots \times \Delta_{j_n}$ if $j_l = \pm j_{l'}$ for some $l \neq l'$. This condition is interpreted in the literature as discarding the hyperplanes $x_l = x_{l'}$ and $x_l = -x_{l'}$, $l, l' = 1, 2, \ldots, n$, $l \neq l'$, from the domain of integration. (Let us observe that in this case,—unlike to the definition of the original Wiener–Itô integrals discussed in Chap. 7,—we omitted also the hyperplanes $x_l = -x_{l'}$ and not only the hyperplanes $x_l = x_l$. $l \neq l'$, from the domain of integration.) Property (a) of the functions in $\bar{\mathcal{H}}_G^n$ and property (v) of the random spectral measures imply that $I_G(f) = \overline{I_G(f)}$, i.e. $I_G(f)$ is a real valued random variable for all $f \in \hat{\bar{\mathcal{H}}}_G^n$. The relation

$$EI_G(f) = 0, \quad \text{for } f \in \hat{\bar{\mathcal{H}}}_G^n, \quad n = 1, 2, \ldots \tag{4.3}$$

also holds. Let $\hat{\mathcal{H}}_G^n = \mathcal{H}_G^n \cap \hat{\bar{\mathcal{H}}}_G^n$. If $f \in \hat{\bar{\mathcal{H}}}_G^n$, then $\operatorname{Sym} f \in \hat{\mathcal{H}}_G^n$, and

$$I_G(f) = I_G(\operatorname{Sym} f). \tag{4.4}$$

Relation (4.4) follows immediately from the observation that

$$Z_G(\Delta_{j_1}) \cdots Z_G(\Delta_{j_n}) = Z_G(\Delta_{\pi(j_1)}) \cdots Z_G(\Delta_{\pi(j_n)}) \quad \text{for all } \pi \in \Pi_n.$$

We also claim that

$$EI_G(f)^2 \leq \frac{1}{n!} \|f\|^2 \quad \text{for } f \in \hat{\bar{\mathcal{H}}}_G^n, \tag{4.5}$$

and

$$EI_G(f)^2 = \frac{1}{n!} \|f\|^2 \quad \text{for } f \in \hat{\mathcal{H}}_G^n. \tag{4.6}$$

More generally, we claim that

$$EI_G(f) I_G(h) = \frac{1}{n!} (f, g) = \int f(x_1, \ldots, x_n) \overline{g(x_1, \ldots, x_n)} \, G(dx_1) \ldots G(dx_n) \quad \text{for } f, g \in \hat{\mathcal{H}}_G^n. \tag{4.7}$$

Because of (4.1) and (4.4) it is enough to check (4.7).

Let $\mathcal{D}$ be a regular system of sets in R^ν, $j_1, \ldots, j_n$ and $k_1, \ldots, k_n$ be indices such that $j_l \neq \pm j_{l'}$, $k_l \neq \pm k_{l'}$ if $l \neq l'$. Then

$$EZ_G(\Delta_{j_1}) \cdots Z_G(\Delta_{j_n}) \overline{Z_G(\Delta_{k_1}) \cdots Z_G(\Delta_{k_n})} = \begin{cases} G(\Delta_{j_1}) \cdots G(\Delta_{j_n}) & \text{if } \{j_1, \ldots, j_n\} = \{k_1, \ldots, k_n\}, \\ 0 & \text{otherwise.} \end{cases}$$

To see the last relation one has to observe that the product on the left-hand side can be written as a product of independent random variables because of property (vii) of the random spectral measures. If $\{j_1,\dots,j_n\} \neq \{k_1,\dots,k_n\}$, then there is an index l such that either $j_l \neq \pm k_{l'}$ for all $1 \le l' \le n$, or there exists an index l', $1 \le l' \le n$, such that $j_l = -k_{l'}$. In the first case $Z_G(\Delta_{j_l})$ is independent of the remaining coordinates of the vector $(Z_G(\Delta_{j_1}),\dots,Z_G(\Delta_{j_n}),\overline{Z_G(\Delta_{k_1})},\dots,\overline{Z_G(\Delta_{k_n})})$, and $EZ_G(\Delta_{j_l}) = 0$. Hence the expectation of the investigated product equals zero, as we claimed. If $j_l = -k_{l'}$ with some index l', then a different argument is needed, since $Z_G(\Delta_{j_l})$ and $Z_G(-\Delta_{j_l})$ are not independent. In this case we can state that since $j_p \neq \pm j_l$ if $p \neq l$, and $k_q \neq \pm j_l$ if $q \neq l'$, the vector $(Z_G(\Delta_{j_l}), Z_G(-\Delta_{j_l}))$ is independent of the remaining coordinates of the above random vector. On the other hand, the product $Z_G(\Delta_{j_l})\overline{Z_G(-\Delta_{j_l})}$ has zero expectation, since $EZ_G(\Delta_{j_l})\overline{Z_G(-\Delta_{j_l})} = G(\Delta_{j_l} \cap (-\Delta_{j_l})) = 0$ by property (iii) of the random spectral measures and the relation $\Delta_{j_l} \cap (-\Delta_{j_l}) = \emptyset$. Hence the expectation of the considered product equals zero also in this case. If $\{j_1,\dots,j_n\} = \{k_1,\dots,k_n\}$, then

$$EZ_G(\Delta_{j_1})\cdots Z_G(\Delta_{j_n})\overline{Z_G(\Delta_{k_1})\cdots Z_G(\Delta_{k_n})} = \prod_{l=1}^n EZ_G(\Delta_{j_l})\overline{Z_G(\Delta_{j_l})} = \prod_{l=1}^n G(\Delta_{j_l}).$$

Therefore for two functions $f, g \in \hat{\mathcal{H}}_G^n$ we may assume that they are adapted to the same regular system $\mathcal{D} = \{\Delta_j,\ j = \pm 1,\dots,\pm N\}$, and

$$\begin{aligned} EI_G(f)I_G(g) &= EI_G(f)\overline{I_G(g)} = \left(\frac{1}{n}\right)^2 \sum\sum f(x_{j_1},\dots,x_{j_n})\overline{g(x_{k_1},\dots,x_{k_n})} \\ &\qquad EZ_G(\Delta_{j_1})\cdots Z_G(\Delta_{j_n})\overline{Z_G(\Delta_{k_1})\cdots Z_G(\Delta_{k_n})} \\ &= \left(\frac{1}{n!}\right)^2 \sum f(x_{j_1},\dots,x_{j_n})\overline{g(x_{j_1},\dots,x_{j_n})}G(\Delta_{j_1})\cdots G(\Delta_{j_n})\cdot n! \\ &= \frac{1}{n!}\int f(x_1,\dots,x_n)\overline{g(x_1,\dots,x_n)}G(dx_1)\cdots G(dx_n) = \frac{1}{n!}(f,g). \end{aligned}$$

We claim that Wiener–Itô integrals of different order are uncorrelated. More explicitly, take two functions $f \in \hat{\bar{\mathcal{H}}}_G^n$ and $f' \in \hat{\bar{\mathcal{H}}}_G^{n'}$ such that $n \neq n'$. Then we have

$$EI_G(f)I_G(f') = 0 \quad \text{if } f \in \hat{\bar{\mathcal{H}}}_G^n,\ \ f' \in \hat{\bar{\mathcal{H}}}_G^{n'},\ \text{and } n \neq n'. \tag{4.8}$$

To see this relation observe that a regular system $\mathcal{D}$ can be chosen is such a way that both f and f' are adapted to it. Then a similar, but simpler argument as the previous one shows that

$$EZ_G(\Delta_{j_1})\cdots Z_G(\Delta_{j_n})\overline{Z_G(\Delta_{k_1})\cdots Z_G(\Delta_{k_{n'}})} = 0$$

for all sets of indices $\{j_1,\dots,j_n\}$ and $\{k_1,\dots,k_{n'}\}$ if $n \neq n'$, hence the sum expressing $EI_G(f)I_G(f')$ in this case equals zero.

We extend the definition of Wiener–Itô integrals to a more general class of kernel functions with the help of the following Lemma 4.1. This is a simple result, but unfortunately it contains several small technical details, and this makes its reading unpleasant.

Lemma 4.1. *The class of simple functions* $\hat{\bar{\mathcal{H}}}_G^n$ *is dense in the (real) Hilbert space* $\bar{\mathcal{H}}_G^n$*, and the class of symmetric simple function* $\hat{\mathcal{H}}_G^n$ *is dense in the (real) Hilbert space* $\mathcal{H}_G^n$.

Proof of Lemma 4.1. It is enough to show that $\hat{\bar{\mathcal{H}}}_G^n$ is dense in the Hilbert space $\bar{\mathcal{H}}_G^n$, since the second statement of the lemma follows from it by a standard symmetrization procedure.

First we reduce the result of Lemma 4.1 to a *Statement A* and then to a *Statement B*. Finally we prove *Statement B*. In *Statement A* we claim that the indicator function χ_A of a bounded set $A \in \mathcal{B}^{n\nu}$ such that $A = -A$ can be well approximated by a function of the form $g = \chi_B \in \hat{\bar{\mathcal{H}}}_G^n$, where χ_B is the indicator function of an appropriate set B. Actually we formulate this statement in a more complicated form, because only in such a way can we reduce the statement about the good approximability of a general, possibly complex valued function $f \in \bar{\mathcal{H}}_G^n$ by a function in $g \in \hat{\bar{\mathcal{H}}}_G^n$ to *Statement A*.

Statement A. Let $A \in \mathcal{B}^{n\nu}$ be a bounded, symmetric set, i.e. let $A = -A$. Then for any $\varepsilon > 0$ there is a function $g \in \hat{\bar{\mathcal{H}}}_G^n$ such that $g = \chi_B$ with some set $B \in \mathcal{B}^{n\nu}$, i.e. g is the indicator function of a set B such that the inequality $\|g - \chi_A\| < \varepsilon$ holds with the norm of the space $\bar{\mathcal{H}}_G^n$. (Here χ_A denotes the indicator function of the set A, and we have $\chi_A \in \bar{\mathcal{H}}_G^n$.)

If $\chi_A \in \bar{\mathcal{H}}_G^n$, and A_1 is such a set for which the set A can be written in the form $A = A_1 \cup (-A_1)$, and the sets A_1 and $-A_1$ have a positive distance from each other, i.e. $\rho(A_1, -A_1) = \inf\limits_{x \in A_1,\, y \in -A_1} \rho(x, y) > \delta$, with some $\delta > 0$, where ρ denotes the Euclidean distance in $R^{n\nu}$, then a good approximation of χ_A can be given with such a function $g = \chi_{B \cup (-B)} \in \hat{\bar{\mathcal{H}}}_G^n$ for which the sets B and $-B$ are separated from each other, and the set B is close to A_1. More explicitly, for all $\varepsilon > 0$ there is a set $B \in \mathcal{B}^{n\nu}$ such that $B \subset A_1^{\delta/2} = \{x\colon\ \rho(x, A_1) \le \frac{\delta}{2}\}$, $g = \chi_{B \cup (-B)} \in \hat{\bar{\mathcal{H}}}_G^n$, and $G^n(A_1 \,\Delta\, B) < \frac{\varepsilon}{2}$. Here $A \Delta B$ denotes the symmetric difference of the sets A and B, and G^n is the n-fold direct product of the spectral measure G on the space $R^{n\nu}$. (The above properties of the set B imply that the function $g = \chi_{B \cup (-B)} \in \hat{\bar{\mathcal{H}}}_G^n$ satisfies the relation $\|g - \chi_A\| < \varepsilon$.)

To justify the reduction of Lemma 4.1 to *Statement A* let us observe that if two functions $f_1 \in \bar{\mathcal{H}}_G^n$ and $f_2 \in \bar{\mathcal{H}}_G^n$ can be arbitrarily well approximated by functions from $\hat{\bar{\mathcal{H}}}_G^n$ in the norm of this space, then the same relation holds for any linear combination $c_1 f_1 + c_2 f_2$ with real coefficients c_1 and c_2. (If the functions f_i are approximated by some functions $g_i \in \hat{\bar{\mathcal{H}}}_G^n$, $i = 1, 2$, then we may assume, by applying some refinement of the partitions if it is necessary, that the approximating functions g_1 and g_2 are adapted to the same regular partition.) Hence the proof about the arbitrarily good approximability of a function $f \in \bar{\mathcal{H}}_G^n$ by functions $g \in \hat{\bar{\mathcal{H}}}_G^n$ can be reduced to the proof about the arbitrarily good approximability of its real part $\text{Re } f \in \bar{\mathcal{H}}_G^n$ and its imaginary part $\text{Im } f \in \bar{\mathcal{H}}_G^n$. Moreover, since the real part and imaginary part of the function f can be arbitrarily well approximated by such real or imaginary valued functions from the space $\bar{\mathcal{H}}_G^n$ which take only finitely many values, the desired approximation result can be reduced to the case when f is the indicator function of a set $A \in \mathcal{B}^{n\nu}$ such that $A = -A$ (if f is real valued), or it takes three values, the value i on a set $A_1 \in \mathcal{B}^{n\nu}$, the value $-i$ on the set $-A_1$, and it equals zero on $R^{n\nu} \setminus (A_1 \cup (-A_1))$ (if f is purely imaginary valued). Besides, the inequalities $G^n(A) < \infty$ and $G^n(A_1) < \infty$ hold. We may even assume that A and A_1 are bounded sets, because $G^n(A) = \lim\limits_{K\to\infty} G^n(A \cap [-K, K]^{n\nu})$, and the same argument applies for A_1.

Statement A immediately implies the desired approximation result in the first case when f is the indicator function of a set A such that $A = -A$. In the second case, when such a function f is considered that takes the values $\pm i$ and zero, observe that the sets $A_1 = \{x\colon f(x) = i\}$ and $-A_1 = \{x\colon f(x) = -i\}$ are disjoint. Moreover, we may assume that they have positive distance from each other, because there are such compact sets $K_N \subset A_1$, $N = 1, 2, \dots$, for which $\lim\limits_{N\to\infty} G^n(A \setminus (K_N \cup (-K_N)) = 0$, and the two disjoint compact sets K_N and $-K_N$ have positive distance. This enables us to restrict our attention to the approximation of such functions f for which $A_1 = \{x\colon f(x) = i\} = K_N$, $-A_1 = \{x\colon f(x) = -i\} = -K_N$ with one of the above defined sets K_N with a sufficiently large index N, and the function f disappears on the complement of the set $A_1 \cup (-A_1)$. To get a good approximation in this case, take $A_1 = K_N$ and apply the second part of *Statement A* for the indicator function $\chi_A = \chi_{K_N \cup (-K_N)}$ with the choice $A_1 = K_N$. We get that there exists a function $g = \chi_{B\cup(-B)} \in \hat{\bar{\mathcal{H}}}_G^n$ such that $B \subset A_1^{\delta/2}$ with a number $\delta > 0$ for which the relation $\rho(K_N, -K_N) > \delta$ holds, and $G^n(A_1 \,\Delta\, B) < \frac{\varepsilon}{2}$. Then we define with the help of the above set B the function $\bar{g} \in \hat{\bar{\mathcal{H}}}_G^n$ as $\bar{g}(x) = i$ if $x \in B$, $\bar{g}(x) = -i$ if $x \in -B$ and $\bar{g}(x) = 0$ otherwise. The definition of the function $\bar{g}(\cdot)$ is meaningful, since $B \cap (-B) = \emptyset$, and it yields a sufficiently good approximation of the function $f(\cdot)$.

In the next step we reduce the proof of *Statement A* to the proof of a result called *Statement B*. In this step we show that to prove *Statement A* it is enough to prove the good approximability of some very special (and relatively simple) indicator functions $\chi_B \in \bar{\mathcal{H}}_G^n$ by a function $g \in \hat{\bar{\mathcal{H}}}_G^n$.

Statement B. Let $B = D_1 \times \dots \times D_n$ be the direct product of bounded sets $D_j \in \mathcal{B}^\nu$ such that $D_j \cap (-D_j) = \emptyset$ for all $1 \le j \le n$. Then for all $\varepsilon > 0$ there is a set $F \subset B \cup (-B)$, $F \in \mathcal{B}^{n\nu}$ such that $\chi_F \in \hat{\bar{\mathcal{H}}}_G^n$, and $\|\chi_{B\cup(-B)} - \chi_F\| \le \varepsilon$, with the norm of the space $\bar{\mathcal{H}}_G^n$.

To deduce *Statement A* from *Statement B* let us first remark that we may reduce our attention to such sets A in *Statement A* for which all coordinates of the points in the set A are separated from the origin. More explicitly, we may assume the existence of a number $\eta > 0$ with the property $A \cap K(\eta) = \emptyset$, where $K(\eta) = \bigcup_{j=1}^{n} K_j(\eta)$ with $K_j(\eta) = \{(x_1, \dots, x_n)\colon x_l \in R^\nu,\ l = 1, \dots, n,\ \rho(x_j, 0) \le \eta\}$. To see our right to make such a reduction observe that the relation $G(\{0\}) = 0$ implies that $\lim_{\eta\to 0} G^n(K(\eta)) = 0$, hence $\lim_{\eta\to 0} G^n(A \setminus K(\eta)) = G^n(A)$. At this point we exploited a weakened form of the non-atomic property of the spectral measure G, namely the relation $G(\{0\}) = 0$.

First we formulate a result that we prove somewhat later, and reduce the proof of *Statement A* to that of *Statement B* with its help. We claim that for all numbers $\varepsilon > 0$, $\bar\delta > 0$ and bounded sets $A \in \mathscr{B}^{n\nu}$ such that $A = -A$, and $A \cup K(\eta) = \emptyset$ there is a finite sequence of bounded sets $B_j \in \mathscr{B}^{n\nu}$, $j = \pm 1, \dots, \pm N$, with the following properties. The sets B_j are disjoint, $B_{-j} = -B_j$, $j = \pm 1, \dots, \pm N$, each set B_j can be written in the form $B_j = D_1^{(j)} \times \cdots \times D_n^{(j)}$ with $D_k^{(j)} \in \mathscr{B}^\nu$, and $D_k^{(-j)} \cap (-D_k^{(j)}) = \emptyset$ for all $1 \le j \le N$ and $1 \le k \le n$, the diameter $d(B_j) = \sup\{\rho(x, y)\colon x, y \in B_j\}$ of the sets B_j has the bound $d(B_j) \le \bar\delta$ for all $1 \le j \le N$, and finally the set $B = \bigcup_{j=1}^{N} (B_j \cup B_{-j})$ satisfies the relation $G^n(A \Delta B) \le \varepsilon$.

Indeed, since we can choose $\varepsilon > 0$ arbitrarily small, the above result together with the application of *Statement B* for all functions $\chi_{B_j \cup (-B_j)}$, $1 \le j \le N$, supplies an arbitrarily good approximation of the function χ_A by a function of the form $\sum_{j=1}^{N} \chi_{F_j} \in \hat{\bar{\mathscr{H}}}_G^n$ in the norm of the space $\bar{\mathscr{H}}_G^n$. Moreover, the random variable $\sum_{j=1}^{N} \chi_{F_j} \in \hat{\bar{\mathscr{H}}}_G^n$ agrees with the indicator function of the set $\bigcup_{j=1}^{N} F_j$, since the sets B_j, $j = \pm 1, \dots, \pm N$, are disjoint, and $F_j \subset B_j \cup B_{-j}$.

If the set A can be written in the form $A = A_1 \cup (-A_1)$ such that $\rho(A_1, -A_1) > \delta$, then we can make the same construction with the only modification that this time we demand that the sets B_j satisfy the relation $d(B_j) \le \bar\delta$ with some $\bar\delta < \frac{\delta}{2}$ for all $1 \le j \le N$. We may assume that $A \cap (B_j \cup B_{-j}) \ne \emptyset$ for all indices j, since we can omit those sets $B_j \cup B_{-j}$ which do not have this property. Since $d(B_j) < \frac{\delta}{2}$, a set B_j cannot intersect both A_1 and $-A_1$. By an appropriate indexation of the sets B_j we have $B_j \subset A_1^{\delta/2}$ and $B_{-j} \subset (-A_1)^{\delta/2}$ for all $1 \le j \le N$. Then the set $B = \bigcup_{j=1}^{N} (B_j \cap F_j)$ and the function $g = \chi_{B \cup (-B)}$ satisfy the second part of *Statement A*.

To find a sequence B_j, $j = \pm 1, \dots, \pm N$, for a set A such that $A = -A$, and $A \cup K(\eta) = \emptyset$ with the properties needed in the above argument observe that there is a sequence of finitely many bounded sets B_j of the form $B_j = D_1^{(j)} \times \cdots \times D_n^{(j)}$, $D_l^{(j)} \in \mathscr{B}^\nu$, $1 \le j \le N$ with some $N < \infty$, whose union $B = \bigcup B_j$ satisfies the relation $G^n(A \,\Delta\, B) < \frac{\varepsilon}{2}$. Because of the symmetry property $A = -A$ of the set A we may assume that these sets B_j have such an indexation with both positive and negative integers for which $B_j = -B_{-j}$. We may also demand that $B_j \cap A \neq \emptyset$ for all sets B_j. Besides, we may assume, by dividing the sets $D_l^{(j)}$ appearing in the definition of the sets B_j into smaller sets if this is needed that their diameter $d(D_l^{(j)}) < \max(\frac{\eta}{2}, \frac{\bar{\delta}}{n})$. This implies because of the relation $A \cap K(\eta) = \emptyset$ that $D_l^{(j)} \cap (-D_l^{(j)}) = \emptyset$ for all j and $1 \leq l \leq n$. The above constructed sets B_j may be non-disjoint, but with the help of their appropriate further splitting and a proper indexation of the sets obtained in such a way we get such a partition of the set B which satisfies all conditions we demanded. For the sake of completeness we present a partition of the set B with the properties we need.

Let us first take for all $1 \leq l \leq n$ the following partition of R^ν with the help of the sets $D_l^{(j)}$, $1 \leq j \leq N$. For a fixed number l this partition consists of all sets $\bar{D}_r^{(l)}$ of the form $\bar{D}_r^{(l)} = \bigcap\limits_{1 \leq j \leq N} F_{l,j}^{r(j)}$, where the indices r are sequences $(r(1), \dots, r(N))$ of length N with $r(j) = 1, 2$ or 3, $1 \leq j \leq N$, and $F_{l,j}^{(1)} = D_l^{(j)}$, $F_{l,j}^{(2)} = -D_l^{(j)}$, $F_{l,j}^{(3)} = R^\nu \setminus (D_l^{(j)} \cup (-D_l^{(j)}))$. Then B can be represented as the union of those sets of the form $\bar{D}_{r_1}^{(1)} \times \cdots \times \bar{D}_{r_n}^{(n)}$ which are contained in B.

Proof of Statement B. To prove *Statement B* first we show that for all $\bar{\varepsilon} > 0$ there is a regular system $\mathscr{D} = \{\Delta_l, l = \pm 1, \dots, \pm N\}$ such that all sets D_j and $-D_j$, $1 \leq j \leq n$, appearing in the formulation of *Statement B* can be expressed as the union of some elements Δ_l of $\mathscr{D}$, and $G(\Delta_l) \leq \bar{\varepsilon}$ for all $\Delta_l \in \mathscr{D}$.

In a first step we prove a weakened version of this statement. We show that there is a regular system $\bar{\mathscr{D}} = \{\Delta'_l, l = \pm 1, \dots, \pm N'\}$ such that all sets D_j and $-D_j$ can be expressed as the union of some sets Δ'_l of $\bar{\mathscr{D}}$. But we have no control on the measure $G(\Delta'_l)$ of the elements of this regular system $\bar{\mathscr{D}}$. To get such a regular system we define the sets $\Delta'(\varepsilon_s, 1 \leq |s| \leq n) = D_1^{\varepsilon_1} \cap (-D_1)^{\varepsilon_{-1}} \cap \cdots \cap D_n^{\varepsilon_n} \cap (-D_n)^{\varepsilon_{-n}}$ for all vectors $(\varepsilon_s, 1 \leq |s| \leq n)$ such that $\varepsilon_s = \pm 1$ for all $1 \leq |s| \leq n$, and the vector $(\varepsilon_s, 1 \leq |s| \leq n)$ contains at least one coordinate $+1$, and $D^1 = D$, $D^{-1} = R^\nu \setminus D$ for all sets $D \in \mathscr{B}^\nu$. Then taking an appropriate reindexation of the sets $\Delta'(\varepsilon_s, 1 \leq |s| \leq n)$ we get a regular system $\bar{\mathscr{D}}$ with the desired properties. (In this construction the sets $\Delta'(\varepsilon_s, 1 \leq |s| \leq n)$ are disjoint, and during their reindexation we drop those of them which equal the empty set.) To see that $\bar{\mathscr{D}}$ with a good indexation is a regular system observe that for a set $\Delta'_l = \Delta'(\varepsilon_s, 1 \leq |s| \leq n) \in \bar{\mathscr{D}}$ we have $-\Delta'_l = \Delta'(\varepsilon_{-s}, 1 \leq |s| \leq n) \in \bar{\mathscr{D}}$, and $\Delta'_l \cap (-\Delta'_l) \subset D_j \cap (-D_j) = \emptyset$ with some index $1 \leq j \leq n$. (We had to exclude the possibility $\Delta_l = -\Delta_l$.)

Next we show that by appropriately refining the above regular system $\bar{\mathscr{D}}$ we can get such a regular system $\mathscr{D} = \{\Delta_l, l = \pm 1, \dots, \pm N\}$ which satisfies the additional property $G(\Delta_l) \leq \bar{\varepsilon}$ for all $\Delta_l \in \mathscr{D}$. To show this let us observe that there is a finite partition $\{E_1, \dots, E_l\}$ of $\bigcup\limits_{j=1}^n (D_j \cup (-D_j))$ such that $G(E_j) \leq \bar{\varepsilon}$ for all $1 \leq j \leq l$. Indeed, the closure of $D = \bigcup\limits_{j=1}^n (D_j \cup (-D_j))$ can be covered by open sets $H_i \subset R^\nu$ such that $G(H_i) \le \bar\varepsilon$ for all sets H_i because of the non-atomic property of the measure G, and by the Heyne–Borel theorem this covering can be chosen finite. With the help of these sets H_i we can get a partition $\{E_1, \dots, E_l\}$ of $\bigcup_{j=1}^{n} (D_j \cup (-D_j))$ with the desired properties.

Then we can make the following construction with the help of the above sets E_j and Δ'_l. Take a pair of elements $(\Delta'_l, \Delta'_{-l}) = (\Delta'_l, -\Delta'_l)$, of $\bar{\mathscr{D}}$, and split up the set Δ'_l with the help of the sets E_j to the union of finitely many disjoint sets of the form $\Delta_{l,j} = \Delta'_l \cap E_j$. Then $G(\Delta_{l,j}) < \bar\varepsilon$ for all sets $\Delta_{l,j}$, and we can write the set Δ'_{-l} as the union of the disjoint sets $-\Delta_{l,j}$. By applying this procedure for all pairs $(\Delta'_l, \Delta'_{-l})$ and by reindexing the sets $\Delta_{l,j}$ obtained by this procedure in an appropriate way we get a regular system $\mathscr{D}$ with the desired properties.

To prove *Property B* let us write $B \cup (-B)$ as the union of products of sets of the form $\Delta_{l_1} \times \dots \times \Delta_{l_n}$ with sets $\Delta_{l_j} \in \mathscr{D}$, $1 \le j \le n$. Here such a regular system $\mathscr{D}$ is considered which satisfies the properties demanded at the start of proof of *Statement B*. Let us discard those products for which $l_j = \pm l_{j'}$ for some pair (j, j'), $j \ne j'$. We define the set F about which we claim that it satisfies Property B as the union of the remaining sets $\Delta_{l_1} \times \dots \times \Delta_{l_n}$. Then $\chi_F \in \hat{\mathscr{H}}_G^n$. Hence to prove that *Statement B* holds with this set F if $\bar\varepsilon > 0$ is chosen sufficiently small it is enough to show that the sum of the terms $G(\Delta_{l_1}) \cdots G(\Delta_{l_n})$ for which $l_j = \pm l_{j'}$ with some $j \ne j'$ is less than $n^2 \bar\varepsilon M^{n-1}$, where $M = \max G(D_j \cup (-D_j)) = 2 \max G(D_j)$. To see this observe that for a fixed pair (j, j'), $j \ne j'$, the sum of all products $G(\Delta_{l_1}) \cdots G(\Delta_{l_n})$ such that $l_j = l_{j'}$ can be bounded by $\bar\varepsilon M^{n-1}$, and the same estimate holds if summation is taken for products with the property $l_j = -l_{j'}$. Indeed, each term of this sum can be bounded by $\bar\varepsilon G^{n-1}\left(\prod_{1 \le p \le n,\, p \ne j} \Delta_{l_p} \right)$, and the events whose G^{n-1} measure is considered in the investigated sum are disjoint. Besides, their union is in the product set $\prod_{1 \le p \le n,\, p \ne j} (D_p \cup D_{-p})$, whose measure is bounded by M^{n-1}. Lemma 4.1 is proved. □

As the transformation $I_G(f)$ is a contraction from $\hat{\mathscr{H}}_G^n$ into $L_2(\Omega, \mathscr{A}, P)$, it can uniquely be extended to the closure of $\hat{\mathscr{H}}_G^n$, i.e. to $\bar{\mathscr{H}}_G^n$. (Here $(\Omega, \mathscr{A}, P)$ denotes the probability space where the random spectral measure $Z_G(\cdot)$ is defined.) At this point we exploit that if $f \in \hat{\mathscr{H}}_G^n$, $N = 1, 2, \dots$, is a convergent sequence in the space $\bar{\mathscr{H}}_G^n$, then the sequence of random variables $I_G(f_N)$ is convergent in the space $L_2(\Omega, \mathscr{A}, P)$, since it is a Cauchy sequence. With the help of this fact and Lemma 4.1 we can introduce the definition of Wiener–Itô integrals in the general case when the integral of a function $f \in \bar{\mathscr{H}}_G^n$ is taken.

Definition of Wiener–Itô Integrals. *Given a function $f \in \bar{\mathscr{H}}_G^n$ with a spectral measure G choose a sequence of simple functions $f_N \in \hat{\mathscr{H}}_G^n$, $N = 1, 2, \dots$, which converges to the function f in the space $\bar{\mathscr{H}}_G^n$. Such a sequence exists by Lemma 4.1. The random variables $I_G(f_N)$ converge to a random variable in the L_2-norm of the* *probability space where these random variables are defined, and the limit does not depend on the choice of the sequence f_N converging to f. This enables us to define the n-fold Wiener–Itô integral with kernel function f as*

$$\int f(x_1,\dots,x_n)Z_G(dx_1)\dots Z_G(dx_n) = n!I_G(f) = \lim_{N\to\infty} n!I_G(f_N),$$

where $f_N \in \hat{\hat{\mathcal{H}}}_G^n$, $N = 1,2,\dots$, is a sequence of simple functions converging to the function f in the space $\bar{\mathcal{H}}_G^n$.

The expression $I_G(f)$ is a real valued random variable for all $f \in \bar{\mathcal{H}}_G^n$, and relations (4.3)–(4.8) remain valid for $f, f' \in \bar{\mathcal{H}}_G^n$ or $f \in \mathcal{H}_G^n$ instead of $f, f' \in \hat{\hat{\mathcal{H}}}_G^n$ or $f \in \hat{\mathcal{H}}_G^n$. Relations (4.6), and (4.8) imply that the transformation I_G: $\operatorname{Exp}\mathcal{H}_G \to L_2(\Omega, \mathcal{A}, P)$ is an isometry. We shall show that also the following result holds.

Theorem 4.2. *Let a stationary Gaussian random field be given (discrete or generalized one), and let Z_G denote the random spectral measure adapted to it. If we integrate with respect to this Z_G, then the transformation I_G: $\operatorname{Exp}\mathcal{H}_G \to \mathcal{H}$, where $\mathcal{H}$ denotes the Hilbert space of the square integrable random variables measurable with respect to the σ-algebra generated by the random variables of the random spectral measure Z_G, is unitary. The transformation $(n!)^{1/2}I_G$: $\mathcal{H}_G^n \to \mathcal{H}_n$ is also unitary.*

In the proof of Theorem 4.2 we need an identity whose proof is postponed to the next chapter.

Theorem 4.3 (Itô's Formula). *Let $\varphi_1,\dots,\varphi_m$, $\varphi_j \in \mathcal{H}_G^1$, $1 \le j \le m$, be an orthonormal system in L_G^2. Let some positive integers $j_1,\dots,j_m$ be given, and let $j_1 + \dots + j_m = N$. Define for all $i = 1,\dots,N$ the function g_i as $g_i = \varphi_s$ for $j_1 + \dots + j_{s-1} < i \le j_1 + \dots + j_s$, $1 \le s \le m$. (In particular, $g_i = \varphi_1$ for $0 < i \le j_1$.) Then*

$$\begin{aligned} &H_{j_1}\left(\int \varphi_1(x)Z_G(dx)\right)\cdots H_{j_m}\left(\int \varphi_m(x)Z_G(dx)\right) \\ &\quad = \int g_1(x_1)\cdots g_N(x_N)\, Z_G(dx_1)\cdots Z_G(dx_N) \\ &\quad = \int Sym\,[g_1(x_1)\cdots g_N(x_N)]\, Z_G(dx_1)\cdots Z_G(dx_N). \end{aligned}$$

($H_j(x)$ denotes again the j-th Hermite polynomial with leading coefficient 1.)

Proof of Theorem 4.2. We have already seen that I_G is an isometry. So it remains to show that it is a one to one map from $\operatorname{Exp}\mathcal{H}_G$ to $\mathcal{H}$ and from $\mathcal{H}_G^n$ to $\mathcal{H}_n$.

The onefold integral $I_G(f)$, $f \in \mathcal{H}_G^1$, agrees with the stochastic integral $I(f)$ defined in Chap. 3. Hence $I_G(e^{i(n,x)}) = X(n)$ in the discrete field case, and

$I_G(\tilde{\varphi}) = X(\varphi)$, $\varphi \in \mathcal{S}$, in the generalized field case. Hence $I_G\colon \mathcal{H}_G^1 \to \mathcal{H}_1$ is a unitary transformation. Let $\varphi_1, \varphi_2, \dots$ be a complete orthonormal basis in $\mathcal{H}_G^1$. Then $\xi_j = \int \varphi_j(x)\, Z_G(dx)$, $j = 1, 2, \dots$, is a complete orthonormal basis in $\mathcal{H}_G^1$. Itô's formula implies that for all sets of positive integers $(j_1, \dots, j_m)$ the random variable $H_{j_1}(\xi_1) \cdots H_{j_m}(\xi_m)$ can be written as a $j_1 + \cdots + j_m$-fold Wiener–Itô integral. Therefore Theorem 2.1 implies that the image of $\operatorname{Exp} \mathcal{H}_G$ is the whole space $\mathcal{H}$, and $I_G\colon \operatorname{Exp} \mathcal{H}_G \to \mathcal{H}$ is unitary.

The image of $\mathcal{H}_G^n$ contains $\mathcal{H}_n$b because of Corollary 2.4 and Itó's formula. Since these images are orthogonal for different n, formula (2.1) implies that the image of $\mathcal{H}_G^n$ coincides with $\mathcal{H}_n$. Hence $(n!)^{1/2} I_G\colon \mathcal{H}_G^n \to \mathcal{H}_n$ is a unitary transformation. □

The next result describes the action of shift transformations in $\mathcal{H}$. We know by Theorem 4.2 that all $\eta \in \mathcal{H}$ can be written in the form

$$\eta = f_0 + \sum_{n=1}^{\infty} \frac{1}{n!} \int f_n(x_1, \dots, x_n) Z_G(dx_1) \dots Z_G(dx_n) \tag{4.9}$$

with $f = (f_0, f_1, \dots) \in \operatorname{Exp} \mathcal{H}_G$ in a unique way, where Z_G is the random measure adapted to the stationary Gaussian field.

Theorem 4.4. *Let $\eta \in \mathcal{H}$ have the form (4.9). Then*

$$T_t \eta = f_0 + \sum_{n=1}^{\infty} \frac{1}{n!} \int e^{i(t, x_1 + \dots + x_n)} f_n(x_1, \dots, x_n) Z_G(dx_1) \dots Z_G(dx_n)$$

for all $t \in R^\nu$ in the generalized field and for all $t \in \mathbb{Z}_\nu$ in the discrete field case.

Proof of Theorem 4.4. Because of formulas (3.9) and (3.10) and the definition of the shift operator T_t we have

$$T_t \left(\int e^{i(n,x)} Z_G(dx) \right) = T_t X_n = X_{n+t} = \int e^{i(t,x)} e^{i(n,x)} Z_G(dx), \quad t \in \mathbb{Z}_\nu,$$

and because of the identity $\widetilde{T_t \varphi}(x) = \int e^{(i(u,x)} \varphi(u - t)\, du = e^{i(t,x)} \tilde{\varphi}(x)$ for $\varphi \in \mathcal{S}$

$$\begin{aligned} T_t \left(\int \tilde{\varphi}(x)\, Z_G(dx) \right) &= T_t X(\varphi) = X(T_t \varphi) \\ &= \int e^{i(t,x)} \tilde{\varphi}(x) Z_G(dx), \quad \varphi \in \mathcal{S}, \quad t \in R^\nu, \end{aligned}$$

in the discrete and generalized field cases respectively. Hence

$$T_t \left(\int f(x) Z_G(dx) \right) = \int e^{i(t,x)} f(x) Z_G(dx) \quad \text{if } f \in \mathcal{H}_G^1$$

for all $t \in \mathbb{Z}_\nu$ in the discrete field and for all $t \in R^\nu$ in the generalized field case. This means that Theorem 4.4 holds in the special case when η is a onefold Wiener–Itô integral. Let $f_1(x), \dots, f_m(x)$ be an orthogonal system in $\mathcal{H}_G^1$. The set of functions $e^{i(t,x)} f_1(x), \dots, e^{i(t,x)} f_m(x)$ is also an orthogonal system in $\mathcal{H}_G^1$. ($t \in \mathbb{Z}_\nu$ in the discrete and $t \in R^\nu$ in the generalized field case.) Hence Itô's formula implies that Theorem 4.4 also holds for random variables of the form

$$\eta = H_{j_1}\left(\int f_1(x) Z_G(dx)\right) \cdots H_{j_m}\left(\int f_m(x) Z_G(dx)\right)$$

and for their finite linear combinations. Since these linear combinations are dense in $\mathcal{H}$ Theorem 4.4 holds true. □

The next result is a formula for the change of variables in Wiener–Itô integrals.

Theorem 4.5. *Let G and G' be two non-atomic spectral measures such that G is absolutely continuous with respect to G', and let $g(x)$ be a complex valued function such that*

$$g(x) = \overline{g(-x)},$$
$$|g^2(x)| = \frac{dG(x)}{dG'(x)}.$$

For every $f = (f_0, f_1, \dots) \in \operatorname{Exp} \mathcal{H}_G$, we define

$$f'_n(x_1, \dots, x_n) = f_n(x_1, \dots, x_n) g(x_1) \cdots g(x_n), \quad n = 1, 2, \dots, \qquad f'_0 = f_0.$$

Then $f' = (f'_0, f'_1, \dots) \in \operatorname{Exp} \mathcal{H}_{G'}^n$, and

$$f_0 + \sum_{n=1}^{\infty} \int \frac{1}{n!} f_n(x_1, \dots, x_n) Z_G(dx_1) \dots Z_G(dx_n)$$
$$\stackrel{\Delta}{=} f'_0 + \sum_{n=1}^{\infty} \frac{1}{n!} \int f'_n(x_1, \dots, x_n) Z_{G'}(dx_1) \dots Z_{G'}(dx_n),$$

where Z_G and $Z_{G'}$ are Gaussian random spectral measures corresponding to G and G'.

Proof of Theorem 4.5. We have $\|f'_n\|_{G'} = \|f_n\|_G$, hence $f' \in \operatorname{Exp} \mathcal{H}_{G'}$. Let $\varphi_1, \varphi_2, \dots$ be a complete orthonormal system in $\mathcal{H}_G^1$. Then $\varphi'_1, \varphi'_2, \dots$, $\varphi'_j(x) = \varphi_j(x) g(x)$ for all $j = 1, 2, \dots$ is a complete orthonormal system in $\mathcal{H}_{G'}^1$. All functions $f_n \in \mathcal{H}_G^n$ can be written in the form $f(x_1, \dots, x_n) = \sum c_{j_1,\dots,j_n} \operatorname{Sym}(\varphi_{j_1}(x_1) \cdots \varphi_{j_n}(x_n))$. Then $f'(x_1, \dots, x_n) = \sum c_{j_1,\dots,j_n} \operatorname{Sym}(\varphi'_{j_1}(x_1) \cdots \varphi'_{j_n}(x_n))$. Rewriting all terms

$$\int \mathrm{Sym}\,(\varphi_{j_1}(x_1)\cdots\varphi_{j_n}(x_n))Z_G(dx_1)\dots Z_G(,dx_n)$$

and

$$\int \mathrm{Sym}\,(\varphi'_{j_1}(x_1)\cdots\varphi'_{j_n}(x_n))Z_{G'}(dx_1)\dots Z_{G'}(,dx_n)$$

by means of Itô's formula we get that f and f' depend on a sequence of independent standard normal random variables in the same way. Theorem 4.5 is proved. □

For the sake of completeness I present in the next Lemma 4.6 another type of change of variable result. I formulate it only in that simple case in which we need it in some later calculations.

Lemma 4.6. *Define for all $t > 0$ the (multiplication) transformation $T_t x = tx$ either from R^ν to R^ν or from the torus $[-\pi,\pi)^\nu$ to the torus $[-t\pi,t\pi)^\nu$. Given a spectral measure G on R^ν or on $[-\pi,\pi)^\nu$ define the spectral measure G_t on R^ν or on $[-t\pi,t\pi)^\nu$ by the formula $G_t(A) = G(\frac{A}{t})$ for all measurable sets A, and similarly define the function $f_{k,t}(x_1,\dots,x_k) = f_k(tx_1,\dots,tx_k)$ for all measurable functions f_k of k variables, $k = 1,2,\dots$, with $x_j \in R^\nu$ or $x_j \in [-\pi,\pi)^\nu$ for all $1 \le j \le k$, and put $f_{0,t} = f_0$. If $f = (f_0,f_1,\dots) \in \mathrm{Exp}\,\mathcal{H}_G$, then $f_t = (f_{0,t},f_{1,t},\dots) \in \mathrm{Exp}\,\mathcal{H}_{G_t}$, and*

$$f_0 + \sum_{n=1}^\infty \int \frac{1}{n!} f_n(x_1,\dots,x_n)Z_G(dx_1)\dots Z_G(dx_n)$$

$$\stackrel{\Delta}{=} f_{0,t} + \sum_{n=1}^\infty \frac{1}{n!}\int f_{n,t}(x_1,\dots,x_n)Z_{G_t}(dx_1)\dots Z_{G_t}(dx_n),$$

where Z_G and Z_{G_t} are Gaussian random spectral measures corresponding to G and G'.

Proof of Lemma 4.6. It is easy to see that $f_t = (f_{0,t},f_{1,t},\dots) \in \mathrm{Exp}\,\mathcal{H}_{G_t}$. Moreover, we may define the random spectral measure Z_{G_t} in the identity we want to prove by the formula $Z_{G_t}(A) = Z_G(\frac{A}{t})$. But with such a choice of Z_{G_t} we can write even $=$ instead of $\stackrel{\Delta}{=}$ in this formula. □

The next result shows a relation between Wick polynomials and Wiener–Itô integrals.

Theorem 4.7. *Let a stationary Gaussian field be given, and let Z_G denote the random spectral measure adapted to it. Let $P(x_1,\dots,x_m) = \sum c_{j_1,\dots,j_n} x_{j_1}\cdots x_{j_n}$ be a homogeneous polynomial of degree n, and let $h_1,\dots,h_m \in \mathcal{H}_G^1$. (Here $j_1,\dots,j_n$ are n indices such that $1 \le j_l \le m$ for all $1 \le l \le n$. It is possible that $j_l = j_{l'}$ also if $l \ne l'$.) Define the random variables $\xi_j = \int h_j(x)Z_G(dx)$,*

$j = 1, 2, \dots, m$, and the function $\tilde{P}(u_1, \dots, u_n) = \sum c_{j_1,\dots,j_n} h_{j_1}(u_1) \cdots h_{j_n}(u_n)$. Then

$$:P(\xi_1, \dots, \xi_m): = \int \tilde{P}(u_1, \dots, u_n) Z_G(du_1) \dots Z_G(du_n).$$

Remark. If P is a polynomial of degree n, then it can be written as $P = P_1 + P_2$, where P_1 is a homogeneous polynomial of degree n, and P_2 is a polynomial of degree less than n. Obviously,

$$:P(\xi_1, \dots, \xi_m): = :P_1(\xi_1, \dots, \xi_m):$$

Proof of Theorem 4.7. It is enough to show that

$$:\xi_{j_1} \cdots \xi_{j_n}: = \int h_{j_1}(u_1) \cdots h_{j_n}(u_n) Z_G(du_1) \dots Z_G(du_n).$$

If $h_1, \dots, h_m \in \mathcal{H}_G^1$ are orthonormal, (all functions h_l have norm 1, and if $l \neq l'$, then h_l and $h_{l'}$ are either orthogonal or $h_l = h_{l'}$), then this relation follows from a comparison of Corollary 2.3 with Itô's formula. In the general case an orthonormal system $\bar{h}_1, \dots, \bar{h}_m$ can be found such that

$$h_j = \sum_{k=1}^{m} c_{j,k} \bar{h}_k, \quad j = 1, \dots, m$$

with some real constants $c_{j,k}$. Set $\eta_k = \int \bar{h}_j Z_G(dx)$. Then

$$\begin{aligned}
:\xi_{j_1} \cdots \xi_{j_n}: &= :\left(\sum_{k=1}^{m} c_{j_1,k} \eta_k\right) \cdots \left(\sum_{k=1}^{m} c_{j_n,k} \eta_k\right): \\
&= \sum_{k_1,\dots,k_n} c_{j_1,k_1} \cdots c_{j_n,k_n} :\eta_{k_1} \cdots \eta_{k_n}: \\
&= \sum_{k_1,\dots,k_n} c_{j_1,k_1} \cdots c_{j_n,k_n} \int \bar{h}_{k_1}(u_1) \cdots \bar{h}_{k_n}(u_n) Z_G(du_1) \dots Z_G(du_n) \\
&= \int h_{j_1}(u_1) \cdots h_{j_n}(u_n) Z_G(du_1) \dots Z_G(du_n)
\end{aligned}$$

as we claimed. □

We finish this chapter by showing how the Wiener–Itô integral can be defined if the spectral measure G may have atoms. We do this although such a construction seems to have a limited importance as in most applications the restriction that we apply the Wiener–Itô integral only in the case of a non-atomic spectral measure G

causes no serious problem. If we try to give this definition by modifying the original one, then we have to split up the atoms. The simplest way we found for this splitting up, was the use of randomization.

Let G be a spectral measure on R^ν, and let Z_G be a corresponding Gaussian spectral random measure on a probability space $(\Omega, \mathcal{A}, P)$. Let us define a new spectral measure $\hat{G} = G \times \lambda_{[-\frac{1}{2},\frac{1}{2}]}$ on $R^{\nu+1}$, where $\lambda_{[-\frac{1}{2},\frac{1}{2}]}$ denotes the uniform distribution on the interval $[-\frac{1}{2}, \frac{1}{2}]$. If the probability space $(\Omega, \mathcal{A}, P)$ is sufficiently rich, a random spectral measure $Z_{\hat{G}}$ corresponding to $\hat{G}$ can be defined on it in such a way that $Z_{\hat{G}}(A \times [-\frac{1}{2}, \frac{1}{2}]) = Z_G(A)$ for all $A \in \mathcal{B}^\nu$. For $f \in \bar{\mathcal{H}}_G^n$ we define the function $\hat{f} \in \bar{\mathcal{H}}_{\hat{G}}^n$ by the formula $\hat{f}(y_1, \dots, y_n) = f(x_1, \dots, x_n)$ if y_j is the juxtaposition (x_j, u_j), $x_j \in R^\nu$, $u_j \in R^1$, $j = 1, 2, \dots, n$. Finally we define the Wiener–Itô integral in the general case by the formula

$$\int f(x_1, \dots, x_n) Z_G(dx_1) \dots Z_G(dx_n) = \int \hat{f}(y_1, \dots, y_n) Z_{\hat{G}}(dy_1) \dots Z_{\hat{G}}(dy_n).$$

(What we actually have done was to introduce a virtual new coordinate u. With the help of this new coordinate we could reduce the general case to the special case when G is non-atomic.) If G is a non-atomic spectral measure, then the new definition of Wiener–Itô integrals coincides with the original one. It is easy to check this fact for onefold integrals, and then Itô's formula proves it for multiple integrals. It can be seen with the help of Itô's formula again, that all results of this chapter remain valid for the new definition of Wiener–Itô integrals. In particular, we formulate the following result.

Given a stationary Gaussian field let Z_G be the random spectral measure adapted to it. All $f \in \mathcal{H}_G^n$ can be written in the form

$$f(x_1, \dots, x_n) = \sum c_{j_1,\dots,j_n} \varphi_{j_1}(x_1) \cdots \varphi_{j_n}(x_n) \tag{4.10}$$

with some functions $\varphi_j \in \mathcal{H}_G^1$, $j = 1, 2, \dots$. Define $\xi_j = \int \varphi_j(x) Z_G(dx)$. If f has the form (4.10), then

$$\int f(x_1, \dots, x_n) Z_G(dx_1) \dots Z_G(dx_n) = \sum c_{j_1,\dots,j_n} :\xi_{j_1} \cdots \xi_{j_n}: .$$

The last identity would provide another possibility for defining Wiener–Itô integrals also in the case when the spectral measure G may have atoms.

Chapter 5
The Proof of Itô's Formula: The Diagram Formula and Some of Its Consequences

We shall prove Itô's formula with the help of the following

Proposition 5.1. *Let $f \in \bar{\mathcal{H}}_G^n$ and $h \in \bar{\mathcal{H}}_G^1$. Let us define the functions*

$$f \underset{k}{\times} h(x_1,\dots,x_{k-1},x_{k+1},\dots,x_n) = \int f(x_1,\dots,x_n)\overline{h(x_k)}G(dx_k), \quad k = 1,\dots,n,$$

and

$$fh(x_1,\dots,x_{n+1}) = f(x_1,\dots,x_n)h(x_{n+1}).$$

Then $f \underset{k}{\times} h$, $k = 1,\dots,n$, and fh are in $\bar{\mathcal{H}}_G^{n-1}$ and $\bar{\mathcal{H}}_G^{n+1}$ respectively, and their norms satisfy the inequality $\|f \underset{k}{\times} h\| \le \|f\|\cdot\|h\|$ and $\|fh\| \le \|f\|\cdot\|h\|$. The relation

$$n!I_G(f)I_G(h) = (n+1)!I_G(fh) + \sum_{k=1}^{n}(n-1)!I_G(f \underset{k}{\times} h)$$

holds true.

We shall get Proposition 5.1 as the special case of the diagram formula formulated in Theorem 5.3.

Remark. There is a small inaccuracy in the formulation of Proposition 5.1. We considered the Wiener–Itô integral of the function $f \underset{k}{\times} h$ with arguments $x_1,\dots$, x_{k-1}, $x_{k+1},\dots$, x_n, while we defined this integral for functions with arguments $x_1,\dots,x_{n-1}$. We can correct this inaccuracy for instance by reindexing the variables of $f \underset{k}{\times} h$ and working with the function

$$(f \underset{k}{\times} h)'(x_1,\dots,x_{n-1}) = f \underset{k}{\times} h(x_{\alpha_k(1)},\dots,x_{\alpha_k(k-1)},x_{\alpha_k(k+1)},\dots,x_{\alpha_k(n)})$$

P. Major, *Multiple Wiener-Itô Integrals*, Lecture Notes in Mathematics 849, DOI 10.1007/978-3-319-02642-8_5,

instead of $f \underset{k}{\times} h$, where $\alpha_k(j) = j$ for $1 \le j \le k-1$, and $\alpha_k(j) = j-1$ for $k+1 \le j \le n$.

We also need the following recursion formula for Hermite polynomials.

Lemma 5.2. *The identity*

$$H_n(x) = xH_{n-1}(x) - (n-1)H_{n-2}(x) \quad \text{for } n = 1, 2, \dots,$$

holds with the notation $H_{-1}(x) \equiv 0$.

Proof of Lemma 5.2.

$$\begin{aligned} H_n(x) = (-1)^n e^{x^2/2} \frac{d^n}{dx^n}(e^{-x^2/2}) &= -e^{x^2/2} \frac{d}{dx}\left(H_{n-1}(x) e^{-x^2/2}\right) \\ &= xH_{n-1}(x) - \frac{d}{dx} H_{n-1}(x). \end{aligned}$$

Since $\frac{d}{dx} H_{n-1}(x)$ is a polynomial of order $n-2$ with leading coefficient $n-1$ we can write

$$\frac{d}{dx} H_{n-1}(x) = (n-1)H_{n-2}(x) + \sum_{j=0}^{n-3} c_j H_j(x).$$

To complete the proof of Lemma 5.2 it remains to show that in the last expansion all coefficients c_j are zero. This follows from the orthogonality of the Hermite polynomials and the calculation

$$\begin{aligned} \int e^{-x^2/2} H_j(x) \frac{d}{dx} H_{n-1}(x)\, dx &= -\int H_{n-1}(x) \frac{d}{dx}(e^{-x^2/2} H_j(x))\, dx \\ &= \int e^{-x^2/2} H_{n-1}(x) P_{j+1}(x)\, dx = 0 \end{aligned}$$

with the polynomial $P_{j+1}(x) = xH_j(x) - \frac{d}{dx} H_j(x)$ of order $j+1$ for $j \le n-3$. □

Proof of Theorem 4.3 via Proposition 5.1. We prove Theorem 4.3 by induction with respect to N. Theorem 4.3 holds for $N = 1$. Assume that it holds for $N-1$. Let us define the functions

$$\begin{aligned} f(x_1, \dots, x_{N-1}) &= g_1(x_1) \cdots g_{N-1}(x_{N-1}) \\ h(x) &= g_N(x). \end{aligned}$$

Then

$$J = \int g_1(x_1)\cdots g_N(x_N) Z_G(dx_1)\dots Z_G(dx_N)$$

$$= N!\, I_G(fh) = (N-1)!\, I_G(f) I_G(h) - \sum_{k=1}^{N-1} (N-2)!\, I_G(f \underset{k}{\times} h)$$

by Proposition 5.1. We can write because of our induction hypothesis that

$$J = H_{j_1}\left(\int \varphi_1(x) Z_G(dx)\right) \cdots H_{j_{m-1}}\left(\int \varphi_{m-1}(x) Z_G(dx)\right)$$
$$H_{j_m-1}\left(\int \varphi_m(x) Z_G(dx)\right) \int \varphi_m(x) Z_G(dx)$$
$$-(j_m-1) H_{j_1}\left(\int \varphi_1(x) Z_G(dx)\right) \cdots H_{j_{m-1}}\left(\int \varphi_{m-1}(x) Z_G(dx)\right)$$
$$H_{j_m-2}\left(\int \varphi_m(x) Z_G(dx)\right),$$

where $H_{j_m-2}(x) = H_{-1}(x) \equiv 0$ if $j_m = 1$. This relation holds, since

$$f \underset{k}{\times} h(x_1,\dots,x_{k-1},x_{k+1},\dots,x_{N-1}) = \int g_1(x_1)\cdots g_{N-1}(x_{N-1}) \overline{\varphi_m(x_k)} G(dx_k)$$

$$= \begin{cases} 0 & \text{if } k \le N-j_m \\ g_1(x_1)\cdots g_{k-1}(x_{k-1}) g_{k+1}(x_{k+1})\cdots g_{N-1}(x_{N-1}) & \text{if } N-j_m < k \le N-1. \end{cases}$$

Hence Lemma 5.2 implies that

$$J = \prod_{s=1}^{m-1} H_{j_s}\left(\int \varphi_s(x) Z_G(dx)\right) \left[H_{j_m-1}\left(\int \varphi_m(x) Z_G(dx)\right) \int \varphi_m(x) Z_G(dx) \right.$$
$$\left. -(j_m-1) H_{j_m-2}\left(\int \varphi_m(x) Z_G(dx)\right)\right] = \prod_{s=1}^{m} H_{j_s}\left(\int \varphi_s(x) Z_G(dx)\right),$$

as claimed. □

Let us fix some functions $h_1 \in \bar{\mathcal{H}}_G^{n_1}, \dots, h_m \in \bar{\mathcal{H}}_G^{n_m}$. In the next result, in the so-called diagram formula, we express the product $n_1! I_G(h_1)\cdots n_m! I_G(h_m)$ as the sum of Wiener–Itô integrals. This result contains Proposition 5.1 as a special case. There is no unique terminology for this result in the literature. We shall follow the notation of Dobrushin in [7]. We introduce a class of diagrams γ denoted by $\Gamma(n_1,\dots,n_k)$ and define with the help of each diagram γ in this class a function h_γ which will be the kernel function of one of the Wiener–Itô integrals taking part in the sum expressing the product of the Wiener–Itô integrals we investigate. First we define the diagrams γ and the functions h_γ corresponding to them, and then we formulate the diagram formula with their help. After the formulation of this result we present an example together with some figures which may help to understand better what the diagram formula is like.

We shall use the term diagram of order $(n_1, \dots, n_m)$ for an undirected graph of $n_1 + \dots + n_m$ vertices which are indexed by the pairs of integers (j, l), $l = 1, \dots, m$, $j = 1, \dots, n_l$, and we shall call the set of vertices (j, l), $1 \le j \le n_l$ the l-th row of the diagram. The diagrams of order $(n_1, \dots, n_m)$ are those undirected graphs with these vertices which have the properties that no more than one edge enters into each vertex, and edges can connect only pairs of vertices from different rows of a diagram, i.e. such vertices (j_1, l_1) and (j_2, l_2) for which $l_1 \ne l_2$. Let $\Gamma = \Gamma(n_1, \dots, n_m)$ denote the set of all diagrams of order $(n_1, \dots, n_m)$.

Given a diagram $\gamma \in \Gamma$ let $|\gamma|$ denote the number of edges in γ. Let there be given a set of functions $h_1 \in \bar{\mathcal{H}}_G^{n_1}, \dots, h_m \in \bar{\mathcal{H}}_G^{n_m}$. Let us denote the variables of the function h_l by $x_{(j,l)}$ instead of x_j, i.e. let us write $h_l(x_{(1,l)}, \dots, x_{(n_l,l)})$ instead of $h_l(x_1, \dots, x_{n_l})$. Put $N = n_1 + \dots + n_m$. We introduce the function of N variables corresponding to the vertices of the diagram by the formula

$$h(x_{(j,l)},\ l = 1, \dots, m,\ j = 1, \dots, n_l) = \prod_{l=1}^{m} h_l(x_{(j,l)},\ j = 1, \dots, n_l). \tag{5.1}$$

For each diagram $\gamma \in \Gamma = \Gamma(n_1, \dots, n_m)$ we define the reenumeration of the indices of the function in (5.1) in the following way. We enumerate the variables $x_{(j,l)}$ in such a way that the vertices into which no edges enter will have the numbers $1, 2, \dots, N - 2|\gamma|$, and the vertices connected by an edge will have the numbers p and $p + |\gamma|$, where $p = N - 2|\gamma| + 1, \dots, N - |\gamma|$. In such a way we have defined a function $h(x_1, \dots, x_N)$ (with an enumeration of the indices of the variables depending on the diagram γ). After the definition of this function $h(x_1, \dots, x_N)$ we take that function of $N - |\gamma|$ variables which we get by replacing the arguments $x_{N-|\gamma|+p}$ by the arguments $-x_{N-2|\gamma|+p}$, $1 \le p \le |\gamma|$ in the function $h(x_1, \dots, x_N)$. Then we define the function h_γ appearing in the diagram formula by integrating this function by the product measure $\prod_{p=1}^{|\gamma|} G(dx_{N-2|\gamma|+p})$.

More explicitly, we write

$$h_\gamma(x_1, \dots, x_{N-2|\gamma|}) = \int \cdots \int h(x_1, \dots, x_{N-|\gamma|}, -x_{N-2|\gamma|+1}, \dots, -x_{N-|\gamma|})$$
$$G(dx_{N-2|\gamma|+1}) \dots G(dx_{N-|\gamma|}). \tag{5.2}$$

The function h_γ depends only on the variables $x_1, \dots, x_{N-2|\gamma|}$, i.e. it is independent of how the vertices connected by edges are indexed. Indeed, it follows from the evenness of the spectral measure that by interchanging the indices s and $s+\gamma$ of two vertices connected by an edge we do not change the value of the integral h_γ. Let us now consider the Wiener–Itô integrals $|\gamma|!I_G(h_\gamma)$. In the diagram formula we shall show that the product of the Wiener–Itô integrals we considered can be expressed as the sum of these Wiener–Itô integrals. To see that the identity appearing in the diagram formula is meaningful observe that although the function h_γ may depend on the numbering of those vertices of γ from which no edge starts, the function $\mathrm{Sym}\, h_\gamma$, and therefore the Wiener–Itô integral $I_G(h_\gamma)$ does not depend on it.

Now I shall formulate the diagram formula. Then I make a remark about the definition of the function h_γ in it and discuss an example to show how to calculate the terms appearing in this result.

Theorem 5.3 (Diagram Formula). *For all functions $h_1 \in \bar{\mathcal{H}}_G^{n_1}, \dots, h_m \in \bar{\mathcal{H}}_G^{n_m}$, $n_1, \dots, n_m = 1, 2, \dots$, the following relations hold:*

(A) $h_\gamma \in \bar{\mathcal{H}}_G^{n-2|\gamma|}$, *and* $\|h_\gamma\| \le \prod_{j=1}^{m} \|h_j\|$ *for all* $\gamma \in \Gamma$.

(B) $n_1! I_G(h_1) \cdots n_m! I_G(h_m) = \sum_{\gamma \in \Gamma} (N - 2|\gamma|)! I_G(h_\gamma)$.

Here $\Gamma = \Gamma(n_1, \dots, n_m)$, and the functions h_γ agree with the functions h_γ defined before the formulation of Theorem 5.3. In particular, h_γ was defined in (5.2).

Remark 1. In the special case $m = 2$, $n_1 = n$, $n_2 = 1$ Theorem 5.3 coincides with Proposition 5.1. To see this it is enough to observe that $h(-x) = \overline{h(x)}$ for all $h \in \bar{\mathcal{H}}_G^1$.

Remark 2. Observe that at the end of the definition of the function h_γ we replaced the variable $x_{N-|\gamma|+p}$ by the variable $-x_{N-2|\gamma|+p}$ and not by $x_{N-2|\gamma|+p}$. This is related to the fact that in the Wiener–Itô integral we integrate with respect a complex valued random measure Z_G which has the property $EZ_G(\Delta)Z_G(-\Delta) = EZ_G(\Delta)\overline{Z_G(\Delta)} = G(\Delta)$, while $EZ_G(\Delta)Z_G(\Delta) = 0$ if $\Delta \cap (-\Delta) = \emptyset$. In the case of the original Wiener–Itô integral considered in Chap. 7 the situation is a bit different. In that case we integrate with respect to a real valued Gaussian orthonormal random measure Z_μ which has the property $EZ_\mu^2(\Delta) = \mu(\Delta)$. In that case a diagram formula also holds, but it has a slightly different form. The main difference is that in that case we define the function h_γ (because of the above mentioned property of the random measure Z_μ) by replacing the variable $x_{N-|\gamma|+p}$ by the variable $x_{N-2|\gamma|+p}$.

To make the notation in the diagram formula more understandable let us consider the following example.

Example. Let us take four functions $h_1 = h_1(x_1, x_2, x_3) \in \bar{\mathcal{H}}_G^3$, $h_2 = h_2(x_1, x_2) \in \bar{\mathcal{H}}_G^2$, $h_3 = h_3(x_1, x_2, x_3, x_4, x_5) \in \bar{\mathcal{H}}_G^5$ and $h_4 = h_4(x_1, x_2, x_3, x_4) \in \bar{\mathcal{H}}_G^4$, and consider the product of Wiener–Itô integrals $3!I_G(h_1)2!I_G(h_2)5!I_G(h_3)4!I_G(h_4)$. Let us look how to calculate the kernel function h_γ of a Wiener–Itô integral $(14 - 2|\gamma|)!I_G(h_\gamma)$, $\gamma \in \Gamma(3, 2, 5, 4)$, appearing in the diagram formula.

We have to consider the class of diagrams $\Gamma(3, 2, 5, 4)$, i.e. the diagrams with vertices which are indexed in the first row as $(1, 1)$, $(2, 1)$, $(3, 1)$, in the second

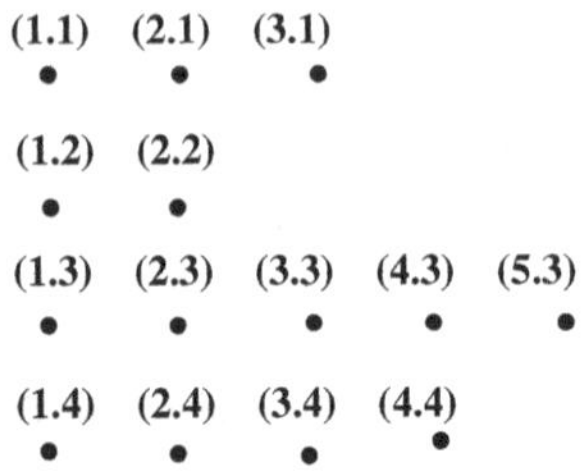

Fig. 5.1 The vertices of the diagrams $\gamma \in \Gamma(3,2,5,4)$

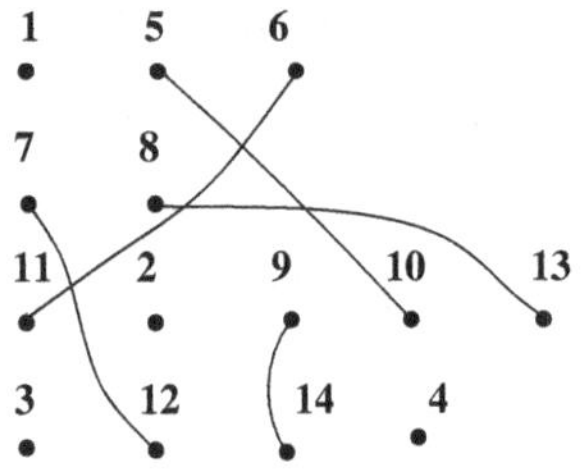

Fig. 5.2 The diagram γ we are working with and the reenumeration of its vertices

row as $(1,2)$, $(2,2)$, in the third row in as $(1,3)$, $(2,3)$, $(3,3)$, $(4,3)$, $(5,3)$ and in the fourth row as $(1,4)$, $(2,4)$, $(3,4)$, $(4,4)$. (See Fig. 5.1.)

Let us take a diagram $\gamma \in \Gamma(3,2,5,4)$, and let us see how we can calculate the kernel function h_γ of the Wiener–Itô integral corresponding to it. We also draw some pictures which may help in following this calculation. Let us consider for instance the diagram $\gamma \in \Gamma(3,2,5,4)$ with edges $((2,1),(4,3))$, $((3,1),(1,3))$, $((1,2),(2,4))$, $((2,2),(5,3))$, $((3,3),(3,4))$. Let us draw the diagram γ with its edges and with such a reenumeration of the vertices which helps in writing up the function $h(\cdot)$ (with $N = 14$ variables) corresponding to this diagram γ and introduced before the definition of the function h_γ.

The function defined in (5.1) equals in the present case

$$h_1(x_{(1,1)}, x_{(2,1)}, x_{(3,1)}) h_2(x_{(1,2)}, x_{(2,2)}) h_3(x_{(1,3)}, x_{(2,3)}, x_{(3,3)}, x_{(4,3)}, x_{(5,3)})$$
$$h_4(x_{(1,4)}, x_{(2,4)}, x_{(3,4)}, x_{(4,4)}).$$

The variables of this function are indexed by the labels of the vertices of γ. We made a relabelling of the vertices of the diagram γ in such a way that by changing the indices of the above function with the help of this relabelling we get the function $h(\cdot)$ corresponding to the diagram γ. In the next step we shall make such a new relabelling of the vertices of γ which helps to write up the functions h_γ we are interested in. (See Fig. 5.2.)

The function $h(\cdot)$ (with $N = 14$ variables) corresponding to the diagram γ can be written (with the help of the labels of the vertices in the second diagram) as

$$h(x_1, x_2, \dots, x_{14})$$
$$= h_1(x_1, x_5, x_6) h_2(x_7, x_8) h_3(x_{11}, x_2, x_9, x_{10}, x_{13}) h_4(x_3, x_{12}, x_{14}, x_4).$$

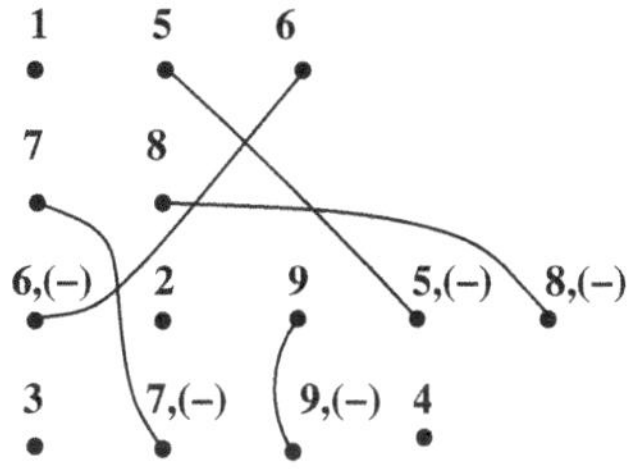

Fig. 5.3 The diagram applied for the calculation of h_γ. The sign $-$ indicates that the corresponding argument is multiplied by -1

Let us change the enumeration of the vertices of the diagram in a way which corresponds to the change of the arguments $x_{N-|\gamma|+p}$ by the arguments $-x_{N-2|\gamma|+p}$. This is done in the next picture. (In this notation the sign (–) denotes that the variable corresponding to this vertex is $-x_{N-2|\gamma|+p}$ and not $x_{N-2|\gamma|+p}$. (See Fig. 5.3.)

With the help of the above diagram we can write up the function

$$h(x_1,\dots,x_{N-|\gamma|},-x_{N-2|\gamma|+1},\dots,-x_{N-|\gamma|})$$

corresponding to the diagram γ in a simple way. This yields that in the present case the function h_γ defined in (5.2) can be written in the form

$$h_\gamma(x_1,x_2,x_3,x_4) = \int\cdots\int h_1(x_1,x_5,x_6)h_2(x_7,x_8)h_3(-x_6,x_2,x_9,-x_5,-x_8)$$
$$h_4(x_3,-x_7,-x_9,x_4)G(\,dx_5)G(\,dx_6)G(\,dx_7)G(\,dx_8)G(\,dx_9).$$

Here we integrate with respect to those variables x_j whose indices correspond to such a vertex of the last diagram from which an edge starts. Then the contribution of the diagram γ to the sum at the right-hand side of diagram formula equals $4!I_G(h_\gamma)$ with this function h_γ.

Let me remark that we had some freedom in choosing the enumeration of the vertices of the diagram γ. Thus e.g. we could have enumerated the four vertices of the diagram from which no edge starts with the numbers 1, 2, 3 and 4 in an arbitrary order. A different indexation of these vertices would lead to a different function h_γ whose Wiener–itô integral is the same. I have chosen that enumeration of the vertices which seemed to be the most natural for me.

Naturally the product of two Wiener–Itô integrals can be similarly calculated, but the notation will be a bit simpler in this case. I briefly show such an example, because in the proof of Theorem 5.3 we shall be mainly interested in the product of two Wiener–Itô integrals.

Example 2. Take two Wiener–Itô integrals with kernel functions $h_1 = h_1(x_1,x_2,x_3) \in \bar{\mathcal{H}}_G^3$ and $h_2 = h_2(x_1,x_2,x_3,x_4,x_5) \in \bar{\mathcal{H}}_G^5$, and calculate the product $3!I_G(h_1)5!I_G(h_2)$ with the help of the diagram formula.

I shall consider only one diagram $\gamma \in \Gamma(3,5)$, and briefly explain how to calculate the kernel function h_γ of the Wiener–Itô integral corresponding to it.

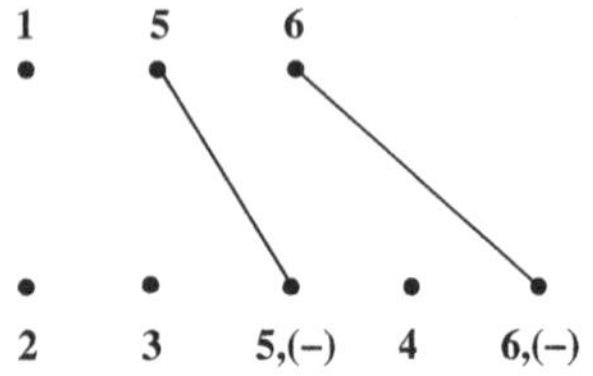

Fig. 5.4 A diagram γ with reenumerated vertices that shows how to calculate the function h_γ

Let us consider for instance the diagram $\gamma \in \Gamma(3,5)$ which contains the edges $((2,1),(3,2))$ and $((3,1),(5,2))$. Then the same calculation as before leads to the introduction of the diagram (Fig. 5.4).

This picture yields the following definition of the diagram h_γ in the present case.

$$h_\gamma(x_1,x_2,x_3,x_4) = \iint h_1(x_1,x_5,x_6)h_2(x_2,x_3,-x_5,x_4,-x_6)G(\,dx_5)G(\,dx_6).$$

Proof of Theorem 5.3. It suffices to prove Theorem 5.3 in the special case $m = 2$. Then the case $m > 2$ follows by induction.

We shall use the notation $n_1 = n$, $n_2 = m$, and we write $x_1,\dots,x_{n+m}$ instead of $x_{(1,1)},\dots,x_{(n,1)},x_{(1,2)}\dots,x_{(m,2)}$. It is clear that the function h_γ satisfies Property (a) of the classes $\bar{\mathscr{H}}_G^{n+m-2|\gamma|}$ defined in Chap. 4. We show that Part (A) of Theorem 5.3 is a consequence of the Schwartz inequality. The validity of this inequality means in particular that the functions h_γ satisfy also Property (b) of the class of functions $\bar{\mathscr{H}}_G^{n+m-2|\gamma|}$.

To prove this estimate on the norm of h_γ it is enough to restrict ourselves to such diagrams γ in which the vertices $(n,1)$ and $(m,2)$, $(n-1,1)$ and $(m-1,2)$,..., $(n-k,1)$ and $(m-k,2)$ are connected by edges with some $0 \le k \le \min(n,m)$. In this case we can write

$$\begin{aligned}
&|h_\gamma(x_1,\dots,x_{n-k-1},x_{n+1},\dots,x_{n+m-k-1})|^2 \\
&\quad = \left| \int h_1(x_1,\dots,x_n)h_2(x_{n+1},\dots,x_{n+m-k-1},-x_{n-k},\dots,-x_n) \right. \\
&\qquad\qquad \left. G(\,dx_{n-k})\dots G(\,dx_n) \right|^2 \\
&\quad \le \int |h_1(x_1,\dots,x_n)|^2 G(\,dx_{n-k})\dots G(\,dx_n) \\
&\qquad\qquad \int |h_2(x_{n+1},\dots,x_{n+m})|^2 G(\,dx_{n+m-k})\dots G(\,dx_{n+m})
\end{aligned}$$

by the Schwartz inequality and the symmetry $G(-A) = G(A)$ of the spectral measure G. Integrating this inequality with respect to the free variables we get Part (A) of Theorem 5.3.

In the proof of Part (B) first we restrict ourselves to the case when $h_1 \in \hat{\bar{\mathcal{H}}}_G^n$ and $h_2 \in \hat{\bar{\mathcal{H}}}_G^m$. Assume that they are adapted to a regular system $\mathcal{D} = \{\Delta_j,\ j = \pm 1, \dots, \pm N\}$ of subsets of R^n with finite measure $G(\Delta_j)$, $j = \pm 1, \dots, \pm N$. We may even assume that all $\Delta_j \in \mathcal{D}$ satisfy the inequality $G(\Delta_j) < \varepsilon$ with some $\varepsilon > 0$ to be chosen later, because otherwise we could split up the sets Δ_j into smaller ones. Let us fix a point $u_j \in \Delta_j$ in all sets $\Delta_j \in \mathcal{D}$. Put $K_i = \sup_x |h_i(x)|$, $i = 1, 2$, and let A be a cube containing all Δ_j.

We can write

$$I = n! I_G(h_1) m! I_G(h_2) = \sum{}' h_1(u_{j_1}, \dots, u_{j_n}) h_2(u_{k_1}, \dots, u_{k_m})$$
$$Z_G(\Delta_{j_1}) \cdots Z_G(\Delta_{j_n}) Z_G(\Delta_{k_1}) \cdots Z_G(\Delta_{k_m})$$

with the numbers $u_{j_p} \in \Delta_{j_p}$ and $u_{k_r} \in \Delta_{k_r}$ we have fixed, where the summation in $\sum'$ goes through all pairs $((j_1, \dots, j_n), (k_1, \dots, k_m))$, $j_p, k_r \in \{\pm 1, \dots, \pm N\}$, $p = 1, \dots, n$, $r = 1, \dots, m$, such that $j_p \neq \pm j_{\bar{p}}$ and $k_r \neq \pm k_{\bar{r}}$ if $p \neq \bar{p}$ or $r \neq \bar{r}$.

Write

$$I = \sum_{\gamma \in \Gamma} \sum{}^{\gamma} h_1(u_{j_1}, \dots, u_{j_n}) h_2(u_{k_1}, \dots, u_{k_m})$$
$$Z_G(\Delta_{j_1}) \cdots Z_G(\Delta_{j_n}) Z_G(\Delta_{k_1}) \cdots Z_G(\Delta_{k_m}),$$

where $\sum^{\gamma}$ contains those terms of $\sum'$ for which $j_p = k_r$ or $j_p = -k_r$ if the vertices $(1, p)$ and $(2, r)$ are connected in γ, and $j_p \neq \pm k_r$ if $(1, p)$ and $(2, r)$ are not connected. Let us define the sets

$$A_1 = A_1(\gamma) = \{p\colon\ p \in \{1, \dots, n\},\ \text{and no edge starts from } (p, 1) \text{ in } \gamma\},$$
$$A_2 = A_2(\gamma) = \{r\colon\ r \in \{1, \dots, m\},\ \text{and no edge starts from } (r, 2) \text{ in } \gamma\}$$

and

$$B = B(\gamma) = \{(p, r)\colon\ p \in \{1, \dots, n\},\ r \in \{1, \dots, m\},$$
$$(p, 1) \text{ and } (r, 2) \text{ are connected in } \gamma\}$$

together with the map $\alpha\colon \{1, \dots, n\} \setminus A_1 \to \{1, \dots, m\} \setminus A_2$ defined as

$$\alpha(p) = r \quad \text{if } (p, r) \in B \quad \text{for all } p \in \{1, \dots, n\} \setminus A_1. \tag{5.3}$$

Let Σ^{γ} denote the value of the inner sum $\sum^{\gamma}$ for some $\gamma \in \Gamma$ in the last summation formula, and write it in the form

$$\Sigma^{\gamma} = \Sigma_1^{\gamma} + \Sigma_2^{\gamma}$$

with

$$\Sigma_1^\gamma = \sum\nolimits^{\gamma} h_1(u_{j_1},\dots,u_{j_n})h_2(u_{k_1},\dots,u_{k_m}) \prod_{p\in A_1} Z_G(\Delta_{j_p}) \prod_{r\in A_2} Z_G(\Delta_{k_r})$$
$$\cdot \prod_{(p,r)\in B} E\left(Z_G(\Delta_{j_p})Z_G(\Delta_{k_r})\right)$$

and

$$\Sigma_2^\gamma = \sum\nolimits^{\gamma} h_1(u_{j_1},\dots,u_{j_n})h_2(u_{k_1},\dots,u_{k_m}) \prod_{p\in A_1} Z_G(\Delta_{j_p}) \prod_{r\in A_2} Z_G(\Delta_{k_r})$$
$$\cdot\left[\prod_{(p,r)\in B} Z_G(\Delta_{j_p})Z_G(\Delta_{k_r}) - E\left(\prod_{(p,r)\in B} Z_G(\Delta_{j_p})Z_G(\Delta_{k_r})\right)\right].$$

The random variables Σ_1^γ and Σ_2^γ are real valued. To see this observe that if the sum defining these expressions contains a term with arguments Δ_{j_p}, and Δ_{k_r}, then it also contains the term with arguments $-\Delta_{j_p}$ and $-\Delta_{k_r}$. This fact together with property (v) of the random spectral measure Z_G and the analogous property of the functions h_1 and h_2 imply that $\Sigma_1^\gamma = \overline{\Sigma_1^\gamma}$ and $\Sigma_2^\gamma = \overline{\Sigma_2^\gamma}$. Hence these random variables are real valued. As a consequence, we can bound $(n+m-2|\gamma|)!I_G(h_\gamma)-\Sigma_1^\gamma$ and Σ_2^γ by means of an estimation of their second moment.

We are going to show that Σ_1^γ is a good approximation of $(n+m-2|\gamma|)!\,I_G(h_\gamma)$, and Σ_2^γ is negligibly small. This implies that $(n+m-2|\gamma|)!I_G(h_\gamma)$ well approximates Σ^γ. The proofs are based on some simple ideas, but unfortunately their description demands a complicated notation which makes their reading unpleasant.

To estimate $(n+m-2|\gamma|)!I_G(h_\gamma)-\Sigma_1^\gamma$ we rewrite Σ_1^γ as a Wiener–Itô integral which can be well approximated by $(n+m-2|\gamma|)!I_G(h_\gamma)$. To find the kernel function of this Wiener–Itô integral we rewrite the sum defining Σ_1^γ by first fixing the variables u_{j_p}, $p\in A_1$, and u_{k_r}, $r\in A_2$, and summing up by the remaining variables, and after this summing by the variables fixed at the first step. We get that

$$\Sigma_1^\gamma = \sum_{\substack{j_p:\,1\le|j_p|\le N \text{ for all } p\in A_1\\ k_r:\,1\le|k_r|\le N \text{ for all } r\in A_2}} h_{\gamma,1}(j_p,\ p\in A_1,\ k_r,\ r\in A_2)$$
$$\prod_{p\in A_1} Z_G(\Delta_{j_p}) \prod_{r\in A_2} Z_G(\Delta_{k_r}) \tag{5.4}$$

with a function $h_{\gamma,1}$ depending on the arguments j_p, $p\in A_1$, and k_r, $r\in A_2$, with values $j_p,k_r\in\{\pm1,\dots,\pm N\}$ defined with the help another function $h_{\gamma,2}$ described below. The function $h_{\gamma,2}$ also depends on the arguments j_p, $p\in A_1$, and k_r, $r\in A_2$, with values $j_p,k_r\in\{\pm1,\dots,\pm N\}$. More explicitly, formula (5.4) holds with the function $h_{\gamma,1}$ defined as

$$h_{\gamma,1}(j_p,\ p\in A_1,\ k_r,\ r\in A_2)=0 \tag{5.5}$$

if the numbers in the set $\{\pm j_p\colon\ p\in A_1\}\cup\{\pm k_r\colon\ r\in A_2\}$ are not all different, and

$$h_{\gamma,1}(j_p,\ p\in A_1,\ k_r,\ r\in A_2)=h_{\gamma,2}(j_p,\ p\in A_1,\ k_r,\ r\in A_2) \tag{5.6}$$

if all numbers $\pm j_p$, $p\in A_1$, and $\pm k_r$, $r\in A_2$ are different, where we define the function $h_{\gamma,2}(j_p,\ p\in A_1,\ k_r,\ r\in A_2)$ for all sequences j_p, $p\in A_1$ and k_r, $r\in A_2$, with $j_p,k_r\in\{\pm1,\dots,\pm N\}$ (i.e. also in the case when some of the arguments j_p, $p\in A_1$, or k_r, $r\in A_2$, agree) by the formula

$$\begin{aligned}h_{\gamma,2}(j_p,\ p\in A_1,\ k_r,\ r\in A_2)\ &=\sum\nolimits^{\gamma,1} h_1(u_{j_1},\dots,u_{j_n})h_2(u_{k_1},\dots,u_{k_m})\\ &\quad\cdot\prod_{(p,r)\in B} E\left(Z_G(\Delta_{j_p})Z_G(\Delta_{k_r})\right). \end{aligned}\tag{5.7}$$

The sum $\sum^{\gamma,1}$ in formula (5.7) which depends on the arguments j_p, $p\in A_1$, and k_r, $r\in A_2$, is defined in the following way. We sum up for such sequences $(j_1,\dots,j_n)$ and $(k_1,\dots,k_m)$ whose coordinates with $p\in A_1$ and $q\in A_2$ are fixed, and agree with the arguments j_p and k_r of the function $h_{\gamma,2}$ at the left-hand side of (5.7) and whose coordinates with indices $p\in\{1,\dots,n\}\setminus A_1$ and $r\in\{1,\dots,m\}\setminus A_2$ satisfy the following conditions.

Put $C=\{\pm j_p,\ p\in A_1\}\cup\{\pm k_r,\ r\in A_2\}$. We demand that all numbers j_p and k_r with indices $p\in\{1,\dots,n\}\setminus A_1$ and $r\in\{1,\dots,m\}\setminus A_2$ are such that $j_p,k_r\in\{\pm1,\dots,\pm N\}\setminus C$. To formulate the remaining conditions let us write all numbers $r\in\{1,\dots,m\}\setminus A_2$ in the form $r=\alpha(p)$, $p\in\{1,\dots,n\}\setminus A_1$ with the map α defined in (5.3). We also demand that only such sequences appear in the summation whose coordinates $k_r=k_{\alpha(p)}$ satisfy the condition $k_{\alpha(p)}=\pm j_p$ for all $p\in\{1,\dots,n\}\setminus A_1$. Besides, all numbers $\pm j_p$, $p\in\{1,\dots,n\}\setminus A_1$, must be different. The summation in $\sum^{\gamma,1}$ is taken for all such sequences j_p, $p\in\{1,\dots,n\}$ and k_r, $r\in\{1,\dots,m\}$, whose coordinates with $p\in\{1,\dots,n\}\setminus A_1$ and $r\in\{1,\dots,m\}\setminus A_2$ satisfy the above conditions.

Formula (5.7) can be rewritten in a simpler form. To do this let us first observe that the condition $k_{\alpha(p)}=\pm j_p$ can be replaced by the condition $k_{\alpha(p)}=-j_p$ in it, and we can write $G(\Delta_{j_p})$ instead of the term $EZ_G(\Delta_{j_p})Z_G(\Delta_{k_r})$ (with $(p,r)\in B$) in the product at the end of (5.7). This follows from the fact that $EZ_G(\Delta_{j_p})Z_G(\Delta_{k_r})=EZ_G(\Delta_{j_p})^2=0$ if $k_r=j_p$ and $EZ_G(\Delta_{j_p})Z_G(\Delta_{k_r})=EZ_G(\Delta_{j_p}Z_G(-\Delta_{j_p})=G(\Delta_{j_p})$ if $k_r=-j_p$. Besides, the expression in (5.7) does not change if we take summation for all such sequences for which the number j_p with coordinate $p\in\{1,\dots,n\}\setminus A$ takes all possible values $j_p\in\{\pm1,\dots,\pm N\}$, because in such a way we only attach such terms to the sum which equal zero. This follows from the fact that both functions h_1 and h_2 are adapted to the regular system $\mathcal{D}$, hence $h_1(u_{j_1},\dots,u_{j_n})h_2(u_{k_1},\dots,u_{k_m})=0$ if for an index $p\in\{1,\dots,n\}\setminus A_1$ $j_p=\pm j_{p'}$ with some $p\neq p'$ or $j_p=-k_r$ with some

$(p,r) \in B$, and the same relation holds if there exists some $r' \in A_2$ such that $j_p = \pm k_{r'}$.

The above relations enable us to rewrite (5.7) in the following simpler form. Let us define that map α^{-1} on the set $\{1,\dots,m\} \setminus A_2$ which is the inverse of the map α defined in (5.3), i.e. $\alpha^{-1}(r) = p$ if $(p,r) \in B$. With this notation we can write

$$\begin{aligned} &h_{\gamma,2}(j_p,\ p \in A_1,\ k_r,\ r \in A_2) \\ &\quad = \sum_{\substack{j_p,\ p\in\{1,\dots,n\}\setminus A_1, \\ 1\le|j_p|\le N \text{ for all indices } p}} h_1(u_{j_1},\dots,u_{j_n}) h_2(u_{k_r},\ r \in A_2, -u_{j_{\alpha^{-1}(r)}},\ r \in \{1,\dots,m\} \setminus A_2) \\ &\qquad \prod_{p\in\{1,\dots,n\}\setminus A_1} G(\Delta_{j_p}). \end{aligned} \tag{5.8}$$

Formula (5.8) can be rewritten in integral form as

$$\begin{aligned} &h_{\gamma,2}(j_p,\ p \in A_1,\ k_r,\ r \in A_2) \\ &\quad = \int h_1(u_{j_p},\ p \in A_1,\ x_p,\ p \in \{1,\dots,n\} \setminus A_1) \\ &\qquad h_2(u_{k_r},\ r \in A_2,\ -x_{\alpha^{-1}(r)},\ r \in \{1,\dots,m\} \setminus A_2) \prod_{p\in\{1,\dots,n\}\setminus A_1} G(dx_p). \end{aligned} \tag{5.9}$$

We define with the help of $h_{\gamma,1}$ and $h_{\gamma,2}$ two new functions on $R^{(n+m-2|\gamma|)\nu}$ with arguments $x_1,\dots,x_{n+m-2|\gamma|}$. The first one will be the kernel function of the Wiener–Itô integral expressing Σ_1^γ, and the second one will be equal to the function h_γ defined in (5.2). We define these functions in two steps. In the first step we reindex the arguments of the functions $h_{1,\gamma}$ and h_2, γ to get functions depending on sequences $j_1,\dots,j_{n+m-2|\gamma|}$. For this goal we list the elements of the sets A_1 and A_2 as $A_1 = \{p_1,\dots,p_{n-|\gamma|}\}$ with $1 \le p_1 < p_2 < \cdots < p_{n-|\gamma|} \le n$ and $A_2 = \{r_1,\dots,r_{m-|\gamma|}\}$ with $1 \le r_1 < r_2 < \cdots < r_{m-|\gamma|} \le m$, and define the maps $\beta_1\colon A_1 \to \{1,\dots,n-|\gamma|\}$ and $\beta_2\colon A_2 \to \{n-|\gamma|+1,\dots,n+m-2|\gamma|\}$ by the formulas $\beta_1(p_l) = l$ if $1 \le l \le n-\gamma$, $1 \le l \le n-|\gamma|$, and $\beta_2(r_l) = l+n-|\gamma|$, $1 \le l \le m-|\gamma|$, if $n-|\gamma|+1 \le l \le n+m-2|\gamma|$. We define with the help of the maps β_1 and β_2 the functions

$$h_{\gamma,3}(j_1,\dots,j_{n+m-2|\gamma|}) = h_{\gamma,1}(j_{\beta_1(r_1)},\dots,j_{\beta_1(n-|\gamma|)}, k_{\beta_2(1)},\dots,k_{\beta_2(m-|\gamma|)})$$

and

$$h_{\gamma,4}(j_1,\dots,j_{n+m-2|\gamma|}) = h_{\gamma,2}(j_{\beta_1(r_1)},\dots,j_{\beta_1(n-|\gamma|)}, k_{\beta_2(1)},\dots,k_{\beta_2(m-|\gamma|)}),$$

where the arguments of the functions $h_{\gamma,3}$ and $h_{\gamma,4}$ are sequences $j_1,\dots,j_{n+m-2|\gamma|}$ with $j_s \in \{\pm1,\dots,\pm N\}$ for all $1 \le s \le n+m-2|\gamma|$.

With the help of the above functions we define the following functions $h_{\gamma,5}$ and $h_{\gamma,6}$ on $R^{(n+m-2|\gamma|)\nu}$.

$$h_{\gamma,5}(x_1,\dots,x_{n+m-2|\gamma|}) = \begin{cases} h_{\gamma,3}(j_1,\dots,j_{n+m-2|\gamma|}) & \text{if } x_l \in \Delta_{j_l}, \\ \quad \text{for all } 1 \le l \le n+m-2|\gamma| \\ 0 & \text{otherwise,} \end{cases}$$

and

$$h_{\gamma,6}(x_1,\dots,x_{n+m-2|\gamma|}) = \begin{cases} h_{\gamma,4}(j_1,\dots,j_{n+m-2|\gamma|}) & \text{if } x_l \in \Delta_{j_l}, \\ \quad \text{for all } 1 \le l \le n+m-2|\gamma| \\ 0 & \text{otherwise.} \end{cases}$$

It follows from relation (5.5) and the definition of the function $h_{\gamma,5}$ (with the help of the definition of the functions $h_{\gamma,1}$ and $h_{\gamma,3}$) that $h_{\gamma,5} \in \hat{\mathcal{H}}_G^n$, and it is adapted to the regular system $\mathcal{D}$. Then relations (5.4) and the definition of $h_{\gamma,5}$ also imply that $\Sigma_1^\gamma = (n+m-2|\gamma|)! I_G(h_{\gamma,5})$.

On the other hand, I claim that the function h_γ defined in (5.2) satisfies the identity $h_\gamma = h_{\gamma,6}$. At this point we must be a bit more careful, because we did not define the function h_γ in a unique way. The function we get by permuting the indices of its variables could be also considered as the function h_γ. This causes no problem, because we are interested not in the function h_γ itself but in the expression $I_G(h_\gamma)$ which does not change if we reindex the variables of the function h_γ. We shall define the function h_γ with a special (natural) indexation of its variables, and we prove the identity $h_\gamma = h_{\gamma,6}$ for this function.

To prove the desired identity first we recall the definition of that function $h(x_1,\dots,x_{n+m})$ (depending on the diagram γ) which was applied in the definition of h_γ. Here we shall use a notation similar to that applied in the definition of the function $h_{\gamma,6}$.

Put $A_1 = \{p_1,\dots,p_{n-|\gamma|}\}$, $1 \le p_1 < p_2 < \dots < p_{n-|\gamma|}$,

$$\{1,\dots,n\} \setminus A_1 = \{q_1,\dots,q_{|\gamma|}\}, \quad 1 \le q_1 < q_2 < \dots < q_{|\gamma|},$$

and $A_2 = \{r_1,\dots,r_{m-|\gamma|}\}$, $1 \le r_1 < r_2 < \dots < r_{m-|\gamma|}$,

$$\{1,\dots,m\} \setminus A_2 = \{q_1',\dots,q_{|\gamma|}'\}, \quad 1 \le q_1' < q_2' < \dots < q_{|\gamma|}',$$

and define with their help the following functions. Define the function $\beta(\cdot)$ on the set $\{1,\dots,n\}$ as $\beta(k) = s$ if $k = p_s \in A_1$, and $\beta(k) = n+m-2|\gamma|+t$ if $k = l_t \in \{1,\dots,n\} \setminus A_1$. Define similarly the function $\delta(\cdot)$ on the set $\{n+1,\dots,n+m\}$ as $\delta(k) = s+|\gamma|$ if $k-n = q_s \in A_2$. If $k-n = l_t' \in \{1,\dots,m\} \setminus A_2$, then there is an edge $(l_t, l_t') \in B$ of the diagram γ, and we define $\delta(k) = n-|\gamma|+t$ with the index t of the number l_t in this case.

With the help of the above notations we can define the function $h(x_1,\dots,x_{n+m})$ which takes part in the definition of h_γ in formula (5.2) as

$$h(x_1,\dots,x_{n+m}) = h_1(x_{\beta(1)},\dots,x_{\beta(n)})h_2(x_{\delta(n+1)},\dots,x_{\delta(n+m)}).$$

To define the kernel function of the integral in (5.2) observe that the set $\{\delta(n+1),\dots,\delta(n+m)\}$ agrees with the set $\{n-|\gamma|+1,\dots,n+m-2|\gamma|\}\cup\{n+m-|\gamma|+1,\dots,n+m\} = C_1\cup C_2$. Put $\bar\delta(k)=\delta(k)$ if $\delta(k)\in C_1$ and $\bar\delta(k)=\delta(k)-|\gamma|$ if $\delta(k)\in C_2$. Let us also introduce $\varepsilon(j)=1$ if $n-|\gamma+1\le j\le n+m-2|\gamma|$, and $\varepsilon(j)=-1$ if $n+m-2|\gamma|+1\le j\le n+m-|\gamma|$. With such a notation we can write

$$\begin{aligned}&h(x_1,\dots,x_{n+m-|\gamma|},-x_{n+m-2|\gamma|+1},\dots,-x_{n+m-|\gamma|})\\&\quad = h_1(x_{\beta(1)},\dots,x_{\beta(n)})h_2(\varepsilon(\bar\delta(n+1))x_{\bar\delta(n+1)},\dots,\varepsilon(\bar\delta(n+m))x_{\bar\delta(n+m)})\end{aligned}$$

as the kernel function in the integral (5.2) defining the function $h_\gamma(x_1,\dots,x_{n+m-2|\gamma|})$ in the present case.

By formula (5.2) we can calculate the function $h_\gamma(x_1,\dots,x_{n+m-2|\gamma|})$ by integrating the above defined function $h(x_1,\dots,x_{n+m-|\gamma|},-x_{n+m-2|\gamma|+1},\dots,-x_{n+m-|\gamma|})$ with respect to the measure $G(dx_{N-2|\gamma|+1})\dots G(dx_{N-|\gamma|})$. By comparing this formula with the definition of the function $h_{\gamma,2}$ defined in (5.9) together with the definition of the functions $h_{\gamma,4}$ and $h_{\gamma,6}$ with its help one can see that the identity $h_{\gamma,6}=h_\gamma$ holds.

We want to compare $I_G(h_{\gamma,5})$ with $I_G(h_{\gamma,6})$. For this goal we have to understand where the functions $h_{\gamma,5}$ and $h_{\gamma,6}$ differ. These are those points $(x_1,\dots,x_{n+m-2|\gamma|})$ where the function $h_{\gamma,5}$ disappears while the function $h_{\gamma,6}$ does not disappear. Observe that in such points $(x_1,\dots,x_{n+m-2|\gamma|})$ where $x_l\in\Delta_{j_l}$, $1\le l\le n+m-2|\gamma|$, with such indices j_l for which the numbers $\pm j_l$, $1\le l\le n+m-2|\gamma|$, are not all different the function $h_{\gamma,5}$ disappears, while the function $h_{\gamma,6}$ may not disappear. But even the function $h_{\gamma,6}$ disappears if some of the numbers in the set $\{\pm j_1,\dots,\pm j_{n-|\gamma|}\}$ or in the set $\{\pm j_{n-|\gamma|+1},\dots,\pm j_{n+m-2|\gamma|}\}$ agree. This fact together with the identity $h_\gamma = h_{\gamma,6}$ and the relation between the functions $h_{\gamma,5}$ and $h_{\gamma,6}$ (implied by the definition of the function $h_{\gamma,1}$ in formulas (5.5) and (5.6)) yield the identity

$$h_\gamma(x_1,\dots,x_{n+m-2|\gamma|}) = h_{\gamma,5}(x_1,\dots,x_{n+m-2|\gamma|}) + h_{\gamma,7}(x_1,\dots,x_{n+m-2|\gamma|})$$

with

$$\begin{aligned}&h_{\gamma,7}(x_1,\dots,x_{n+m-2|\gamma|})\\&\quad=\begin{cases} h_\gamma(x_1,\dots,x_{n+m-2|\gamma|}) & \text{if there exist indices } j_l,\ 1\le|j_l|\le N,\\ \quad 1\le l\le n+m-2|\gamma| \text{ such that } x_l\in\Delta_{j_l},\ 1\le l\le n+m-2|\gamma|,\\ \quad \text{all numbers } \pm j_1,\dots,\pm j_{n-2|\gamma|} \text{ are different},\\ \quad \text{all numbers } \pm j_{n-|\gamma|+1},\dots,\pm j_{n+m-2|\gamma|} \text{ are different},\\ \quad \text{and } \{\pm j_1,\dots,\pm j_{n-|\gamma|}\}\cap\{\pm j_{n-|\gamma|+1},\dots,\pm j_{n+m-2|\gamma|}\}\ne\emptyset\\ 0 & \text{otherwise.}\end{cases}\end{aligned}$$

Since $\Sigma_1^\gamma = (n+m-2|\gamma|)! I_G(h_{\gamma,5})$, we have

$$(n+m-2|\gamma|)! I_G(h_\gamma) - \Sigma_1^\gamma = (n+m-2|\gamma|)! I_G(h_{\gamma,7}),$$

and

$$E(\Sigma_1^\gamma - (n+m-2|\gamma|)! I_G(h_\gamma))^2 \le (n+m-2|\gamma|)! \|h_{\gamma,7}(\cdot)\|^2$$

with the norm $\|\cdot\|$ in $\bar{\mathcal{H}}_G^{n+m-2|\gamma|}$.

On the other hand,

$$\sup |h_{\gamma,7}(x_1,\dots,x_{n+m-2|\gamma|})| \le \sup |h_\gamma(x_1,\dots,x_{n+m-2|\gamma|})| \le K_1 K_2 L^{|\gamma|},$$

with $K_1 = \sup |h_1|$, $K_2 = \sup |h_2|$, and $L = G(A)$, where A is a fixed cube containing all Δ_j. Hence

$$\begin{aligned} E(\Sigma_1^\gamma - (n+m-2|\gamma|)! I_G(h_\gamma))^2 &\le C_1 \|h_{\gamma,7}\|^2 \\ &\le C_2 \sum\nolimits'' G(\Delta_{j_1}) \cdots G(\Delta_{j_{n+m-2|\gamma|}}) \\ &\le C \sup_j G(\Delta_j) \le C\varepsilon, \end{aligned} \tag{5.10}$$

where the summation $\sum''$ goes for such sequences $j_1,\dots,j_{n+m-2|\gamma|}$, $1 \le |j_l| \le N$ for all $1 \le l \le n+m-2|\gamma|$, for which all numbers $\pm j_1,\dots,\pm j_{n-|\gamma|}$ are different, the same relation holds for the elements of the sequence $\pm j_{n-|\gamma|+1},\dots,\pm j_{n+m-2|\gamma|}$, and

$$\{\pm j_1,\dots,\pm j_{n-|\gamma|}\} \cap \{\pm j_{n-|\gamma|+1},\dots,\pm j_{n+m-2|\gamma|}\} \ne \emptyset.$$

The constants C_1, C_2 and C may depend on the functions h_1, h_2 and spectral measure G, but they do not depend on the regular system $\mathcal{D}$, hence in particular on the parameter ε. In the verification of the last inequality in (5.10) we can exploit that each term in the sum $\sum''$ is a product which contains a factor $G(\Delta_j)^2 \le \varepsilon G(\Delta_j)$. Here an argument can be applied which is similar to the closing step in the proof of Lemma 4.1, to the final argument in the proof of *Statement B*.

Now we turn to the estimation of $E(\Sigma_2^\gamma)^2$. It can be expressed as a linear combination of terms of the form

$$\begin{aligned}&\Sigma_3^\gamma(j_p,k_r,j_{\bar p},k_{\bar r},\; p,\bar p\in\{1,\dots,n\},\; r,\bar r\in\{1,\dots,m\}) \qquad (5.11)\\ &\quad= E\Bigg(\Bigg(\prod_{p\in A_1} Z_G(\Delta_{j_p})\prod_{r\in A_2} Z_G(\Delta_{k_r})\prod_{\bar p\in A_1} Z_G(\Delta_{j_{\bar p}})\prod_{\bar r\in A_2} Z_G(\Delta_{k_{\bar r}})\Bigg)\\ &\qquad\left[\prod_{(p,r)\in B} Z_G(\Delta_{j_p})Z_G(\Delta_{k_r}) - E\prod_{(p,r)\in B} Z_G(\Delta_{j_p})Z_G(\Delta_{k_r})\right]\\ &\qquad\left[\prod_{(\bar p,\bar r)\in B} Z_G(\Delta_{j_{\bar p}})Z_G(\Delta_{k_{\bar r}}) - E\prod_{(\bar p,\bar r)\in B} Z_G(\Delta_{j_{\bar p}})Z_G(\Delta_{k_{\bar r}})\right]\Bigg),\end{aligned}$$

where Σ_3^γ depends on such sequences of numbers j_p, k_r, $j_{\bar p}$, $k_{\bar r}$ with indices $1\le p,\bar p\le n$ and $1\le r,\bar r\le m$ for which $j_p,k_r,j_{\bar p},k_{\bar r}\in\{\pm1,\dots,\pm N\}$ for all indices $p,r,\bar p$ and $\bar r$, $j_p=k_r$ or $j_p=-k_r$ if $(p,r)\in B$, otherwise all numbers $\pm j_p$, $\pm k_r$ are different, and the same relations hold for the indices $j_{\bar p}$ and $k_{\bar r}$ if the indices p and r are replaced by $\bar p$ and $\bar r$. Moreover, the absolute value of all coefficients in this linear combination which depend on the functions h_1 and h_2 is bounded by $\sup|h_1(x)|^2\sup|h_2(x)|^2$.

We want to show that for most sets of arguments $(j_p,\,k_r,\,j_{\bar p},\,k_{\bar r})$ the expression Σ_3^γ equals zero, and it is also small in the remaining cases.

Let us fix a sequence of arguments j_p, k_r, $j_{\bar p}$, $k_{\bar r}$ of Σ_3^γ, and let us estimate its value with these arguments. Define the sets

$$\mathscr{A}=\{j_p\colon\, p\in A_1\}\cup\{k_r\colon\, r\in A_2\}\quad\text{and}\quad \bar{\mathscr{A}}=\{j_{\bar p}\colon\, \bar p\in A_1\}\cup\{k_{\bar r}\colon\, \bar r\in A_2\}.$$

We claim that Σ_3^γ equals zero if $\bar{\mathscr{A}}\neq-\mathscr{A}$.

In this case there exists an index $l\in\mathscr{A}$ such that $-l\notin\bar{\mathscr{A}}$. Let us carry out the multiplication in (5.11). Because of the independence properties of random spectral measures each product in this expression can be written as the product of independent factors, and the independent factor containing the term $Z_G(\Delta_l)$ has zero expectation. To see this observe that the set Δ_l appears exactly once among the arguments of the terms $Z_G(\Delta_{j_p})$ and $Z_G(\Delta_{k_r})$, and none of these terms contains the argument $-\Delta_l=\Delta_{-l}$. Although $-l\notin\bar{\mathscr{A}}$, it may happen that $l\in\bar{\mathscr{A}}$. In this case the product under investigation contains the independent factor $Z_G(\Delta_l)^2$ with $EZ_G(\Delta_l)^2=0$. If $l\notin\bar{\mathscr{A}}$, then there are two possibilities. Either this product contains an independent factor of the form $Z_G(\Delta_l)$ with $EZ_G(\Delta_l)=0$, or there is a pair $(\bar p,\bar r)\in B$ such that $(j_{\bar p},k_{\bar r})=(\pm l,\pm l)$, and an independent factor of the form $Z_G(\Delta_l)Z_G(\pm\Delta_{-l})Z_G(\pm\Delta_l)$ with the property $EZ_G(\Delta_l)Z_G(\pm\Delta_{-l})Z_G(\pm\Delta_l)=0$ appears. Hence $\Sigma_3^\gamma=0$ also in this case.

Let

$$\mathscr{F}=\bigcup_{(p,r)\in B}\{j_p,k_r\}\quad\text{and}\quad \bar{\mathscr{F}}=\bigcup_{(\bar p,\bar r)\in B}\{(j_{\bar p},k_{\bar r}\}.$$

A factorization argument shows again that the expression in (5.11) equals zero if the sets $\mathcal{F} \cup (-\mathcal{F})$ and $\bar{\mathcal{F}} \cup (-\bar{\mathcal{F}})$ are disjoint.

In the proof of this statement we can restrict ourselves to the case when $\mathcal{A} = -\bar{\mathcal{A}}$. In this case $\pm\mathcal{A}$ is disjoint both of $\mathcal{F} \cup (-\mathcal{F})$ and $\bar{\mathcal{F}} \cup (-\bar{\mathcal{F}})$, and the product under investigation contains the independent factor $\prod\limits_{(p,r)\in B} Z_G(\Delta_{j_p})Z_G(\Delta_{k_r}) - E \prod\limits_{(p,r)\in B} Z_G(\Delta_{j_p})Z_G(\Delta_{k_r})$ with expectation zero.

Finally in the remaining cases when $\mathcal{F} \cup (-\mathcal{F})$ and $\bar{\mathcal{F}} \cup (-\bar{\mathcal{F}})$ are not disjoint, and $\mathcal{A} = -\bar{\mathcal{A}}$ the absolute value of the expression in (5.11) can be estimated from above by

$$C\,\varepsilon \prod G(\Delta_{j_p})G(\Delta_{k_r})G(\Delta_{j_{\bar{p}}})G(\Delta_{k_{\bar{r}}}) \tag{5.12}$$

with a universal constant $C < \infty$ depending only on the parameters n and m, where the indices j_p, k_r, $j_{\bar{p}}$, $k_{\bar{r}}$ are the same as in (5.11) with the following difference: All indices appear in (5.12) with multiplicity 1, and if both indices l and $-l$ are present in (5.11), then one of them is omitted form (5.12). Moreover, for all j_p one of terms $G(\Delta_{\pm j_p})$ really appears in this product, and the analogous statement also holds for all indices k_r, $j_{\bar{p}}$ and $k_{\bar{r}}$. The multiplying term ε appears in (5.12), since by carrying out the multiplications in (5.11) and factorizing each term, we get that all non-zero terms have a factor either of the form

$$\begin{aligned} EZ_G(\Delta)^2 Z_G(-\Delta)^2 &= E(\operatorname{Re} Z_G(\Delta)^2 + \operatorname{Im} Z_G(\Delta)^2)^2 \\ &= E \operatorname{Re} Z_G(\Delta)^4 + E \operatorname{Im} Z_G(\Delta)^4 \\ &\quad +2E \operatorname{Re} Z_G(\Delta)^2 E \operatorname{Im} Z_G(\Delta)^2 = 8G(\Delta)^2 \end{aligned}$$

or of the form $\left(E|Z_G(\Delta)|^2\right)^2 = G(\Delta)^2$, and $G(\Delta) < \varepsilon$ for all $\Delta \in \mathcal{D}$. (We did not mention the possibility of an independent factor of the form $EZ_G(\Delta)^4$ or $EZ_G(\Delta)^3 Z_G(-\Delta)$ with $\Delta \in \mathcal{D}$, because as some calculation shows, $EZ_G(\Delta)^4 = 0$ and $EZ_G(\Delta)^3 Z_G(-\Delta) = 0$.)

Let us express $E(\Sigma_2^\gamma)^2$ as the linear combination of the quantities Σ_3^γ, and let us bound each term Σ_3^γ in the above way. This supplies an upper bound for $E(\Sigma_2^\gamma)^2$ by means of a sum of terms of the form (5.12). Moreover, some consideration shows that each of these terms appears only with a multiplicity less than $C(n,m)$ with an appropriate constant $C(n,m)$. Hence we can write

$$E(\Sigma_2^\gamma)^2 \le K_1^2 K_2^2 C(n,m) C\varepsilon \sum_{r=1}^{n+m} \sum_{j_1,\dots,j_r}{}''' G(\Delta_{j_1}) \cdots G(\Delta_{j_r}),$$

where the indices $j_1, \dots, j_r \in \{\pm 1, \dots, \pm N\}$ in the sum $\sum'''$ are all different, and $K_j = \sup |h_j(x)|$, $j = 1, 2$. Hence

$$E(\Sigma_2^\gamma)^2 \le C_1\varepsilon \sum_{r=1}^{n+m} G(A)^r \le C_2\varepsilon$$

with some appropriate constants C_1 and C_2. Because of the inequality (5.10), the identity $n!I_G(h_1)m!I_G(h_2) = \sum_{\gamma\in\Gamma}(\Sigma_1^\gamma + \Sigma_\gamma^2)$ and the last relation the inequality

$$E\left(n!I_G(h_1)m!I_G(h_2) - \sum_{\gamma\in\Gamma}(n+m-2|\gamma|)!I_G(h_\gamma)\right)^2$$
$$= E\left(\sum_{\gamma\in\Gamma}(\Sigma_1^\gamma + \Sigma_2^\gamma - (n+m-2|\gamma|)!\,I_G(h_\gamma))\right)^2$$
$$\le C_3\left(\sum_{\gamma\in\Gamma} E((m+n-2|\gamma|)!\,I_G(h_\gamma) - \Sigma_1^\gamma)^2 + E(\Sigma_2^\gamma)^2\right) \le C_4\varepsilon$$

holds. Since $\varepsilon > 0$ can be chosen arbitrarily small, Part B is proved in the special case $h_1 \in \hat{\bar{\mathcal{H}}}_G^n$, $h_2 \in \hat{\bar{\mathcal{H}}}_G^m$.

If $h_1 \in \bar{\mathcal{H}}_G^n$ and $h_2 \in \bar{\mathcal{H}}_G^m$, then let us choose a sequence of functions $h_{1,r} \in \hat{\bar{\mathcal{H}}}_G^n$ and $h_{2,r} \in \hat{\bar{\mathcal{H}}}_G^m$ such that $h_{1,r} \to h_1$ and $h_{2,r} \to h_2$ in the norm of the spaces $\bar{\mathcal{H}}_G^n$ and $\bar{\mathcal{H}}_G^m$ respectively. Define the functions $\hat{h}_\gamma(r)$ and $h_\gamma(r)$ in the same way as h_γ, but substitute the pair of functions (h_1, h_2) by $(h_{1,r}, h_2)$ and $(h_{1,r}, h_{2,r})$ in their definition. We shall show with the help of Part (A) that

$$E|I_G(h_1)I_G(h_2) - I_G(h_{1,r})I_G(h_{2,r})| \to 0,$$

and

$$E|I_G(h_\gamma) - I_G(h_\gamma(r))| \to 0 \quad \text{for all } \gamma \in \Gamma$$

as $r \to \infty$. Then a simple limiting procedure shows that Theorem 5.3 holds for all $h_1 \in \bar{\mathcal{H}}_G^n$ and $h_2 \in \bar{\mathcal{H}}_G^m$.

We have

$$\begin{aligned}
&E|I_G(h_1)I_G(h_2) - I_G(h_{1,r})I_G(h_{2,r})|\\
&\quad \le E|(I_G(h_1 - h_{1,r}))I_G(h_2)| + E|I_G(h_{1,r})I_G(h_2 - h_{2,r})|\\
&\quad \le \frac{1}{n!m!}\left(\|h_1 - h_{1,r}\|^{1/2}\|h_2\|^{1/2} + \|h_2 - h_{2,r}\|^{1/2}\|h_{1,r}\|\right) \to 0,
\end{aligned}$$

and by Part (A) of Theorem 5.3

$$
\begin{aligned}
E|I_G(h_\gamma) - I_G(h_\gamma(r))| &\le E|I_G(h_\gamma) - I_G(\hat{h}_\gamma(r))| + E|I_G(h_\gamma(r)) - I_G(\hat{h}_\gamma(r))| \\
&\le \|h_\gamma - \hat{h}_\gamma(r)\|^{1/2} + \|h_\gamma(r) - \hat{h}_\gamma(r)\|^{1/2} \\
&\le \|h_1 - \hat{h}_{1,r}\|^{1/2}\|h_2\|^{1/2} + \|h_2 - \hat{h}_{2,r}\|^{1/2}\|h_{1,r}\|^{1/2} \to 0.
\end{aligned}
$$

Theorem 5.3 is proved. □

We formulate some consequences of Theorem 5.3. Let $\bar{\Gamma} \subset \Gamma$ denote the set of complete diagrams, i.e. let a diagram $\gamma \in \bar{\Gamma}$ if an edge enters in each vertex of γ. We have $EI(h_\gamma) = 0$ for all $\gamma \in \Gamma \setminus \bar{\Gamma}$, since (4.3) holds for all $f \in \bar{\mathscr{H}}_G^n$, $n \ge 1$. If $\gamma \in \bar{\Gamma}$, then $I(h_\gamma) \in \bar{\mathscr{H}}_G^0$. Let h_γ denote the value of $I(h_\gamma)$ in this case. Now we have the following

Corollary 5.4. *For all $h_1 \in \bar{\mathscr{H}}_G^{n_1}, \dots, h_n \in \bar{\mathscr{H}}_G^{n_m}$*

$$
En_1! I_G(h_1) \cdots n_m! I_G(h_m) = \sum_{\gamma \in \bar{\Gamma}} h_\gamma .
$$

(The sum on the right-hand side equals zero if $\bar{\Gamma}$ is empty.)

As a consequence of Corollary 5.4 we can calculate the expectation of products of Wick polynomials of Gaussian random variables.

Let $X_{k,j}$, $EX_{k,j} = 0$, $1 \le k \le p$, $1 \le j \le n_k$, be a sequence of (jointly) Gaussian random variables. We want to calculate the expected value of the product of the Wick polynomials $:X_{k,1} \cdots X_{k,n_k}:$, $1 \le k \le p$, if we know all covariances $EX_{k,j} X_{\bar{k},\bar{j}} = a((k,j),(\bar{k},\bar{j}))$, $1 \le k, \bar{k}, \le p$, $1 \le j \le n_k$, $1 \le \bar{j} \le \bar{n}_k$. For this goal let us consider the class of closed diagrams $\bar{\Gamma}(k_1, \dots, k_p)$, and define the following quantity $\gamma(A)$ depending on the closed diagrams γ and the set A of all covariances $EX_{k,j} X_{\bar{k},\bar{j}} = a((k,j),(\bar{k},\bar{j}))$

$$
\gamma(A) = \prod_{((k,j),(\bar{k},\bar{j})) \text{ is an edge of } \gamma} a((k,j),(\bar{k},\bar{j})), \qquad \gamma \in \Gamma.
$$

With the above notation we can formulate the following result.

Corollary 5.5. *Let $X_{k,j}$, $EX_{k,j} = 0$, $1 \le k \le p$, $1 \le j \le n_k$, be a sequence of Gaussian random variables. Let $a((k,j),(\bar{k},\bar{j})) = EX_{k,j} X_{\bar{k},\bar{j}}$, $1 \le k, \bar{k}, \le p$, $1 \le j \le n_k$, $1 \le \bar{j} \le \bar{n}_k$ denote the covariances of these random variables. Then the expected value of the product of the Wick polynomials $:X_{k,1} \cdots X_{k,n_k}:$, $1 \le k \le p$, can be expressed as*

$$
E\left(\prod_{k=1}^{p} :X_{k,1} \cdots X_{k,n_k}: \right) = \sum_{\gamma \in \bar{\Gamma}(k_1,\dots,k_p)} \gamma(A)
$$

with the above defined quantities $\gamma(A)$. In the case when $\bar{\Gamma}(k_1,\dots,k_p)$ is empty, e.g. if $k_1+\cdots+k_p$ is an odd number, the above expectation equals zero.

Remark. In the special case when $X_{k,1}=\cdots=X_{k,n_k}=X_k$, and $EX_k^2=1$ for all indices $1\le k\le p$ Corollary 5.5 provides a formula for the expectation of the product of Hermite polynomials of standard normal random variables. In this case we have $a((k,j),(\bar{k},\bar{j}))=\bar{a}(k,\bar{k})$ with a function $\bar{a}(\cdot,\cdot)$ not depending on the arguments j and $\bar{j}$, and the left-hand side of the identity in Corollary 5.5 equals $EH_{n_1}(X_1)\cdots H_{n_p}(X_p)$ with standard normal random variables $X_1,\dots,X_n$ with correlations $EX_kX_{\bar{k}}=\bar{a}(k,\bar{k})$.

Proof of Corollary 5.5. We can represent the random variables $X_{k,j}$ in the form $X_{k,j}=\sum\limits_p c_{k,j,p}\xi_p$ with some appropriate coefficients $c_{k,j,p}$, where $\xi_1,\xi_2,\dots$ is a sequence of independent standard normal random variables. Let $Z(dx)$ denote a random spectral measure corresponding to the one-dimensional spectral measure with density function $g(x)=\frac{1}{2\pi}$ for $|x|<\pi$, and $g(x)=0$ for $|x|\ge\pi$. The random integrals $\int e^{ipx}Z(dx)$, $p=0,\pm1,\pm2,\dots$, are independent standard normal random variables. Define $h_{k,j}(x)=\sum\limits_p c_{k,j,p}e^{ipx}$, $k=1,\dots,p$, $1\le j\le n_k$. The random variables $X_{k,j}$ can be identified with the random integrals $\int h_{k,j}(x)Z(dx)$, $k=1,\dots,p$, $1\le j\le n_k$, since their joint distributions coincide. Put $\hat{h}_k(x_1,\dots,x_{n_k})=\prod\limits_{j=1}^{n_k}h_{k,j}(x_j)$. It follows from Theorem 4.7 that

$$:X_{k,1}\cdots X_{k,n_k}:=\int\hat{h}_k(x_1,\dots,x_{n_k})Z(dx_1)\dots Z(dx_{n_k})=n_k!I(\hat{h}_k(x_1,\dots,x_{n_k}))$$

for all $1\le k\le p$. Hence an application of Corollary 5.4 yields Corollary 5.5. One only has to observe that $\int_{-\pi}^{\pi}h_{k,j}(x)\overline{h_{\bar{k},\bar{j}}(x)}\,dx=a((k,j),(\bar{k},\bar{j}))$ for all $k,k=1,\dots,p$ and $1\le j\le n_k$. □

Theorem 5.3 states in particular that the product of Wiener–Itô integrals with respect to a random spectral measure of a stationary Gaussian fields belongs to the Hilbert space $\mathcal{H}$ defined by this field, since it can be written as a sum of Wiener–Itô integrals. This means a trivial measurability condition, and also that the product has a finite second moment, which is not so trivial. Theorem 5.3 actually gives the following non-trivial inequality.

Let $h_1\in\mathcal{H}_G^{n_1},\dots,h_m\in\mathcal{H}_G^{n_m}$. Let $|\bar{\Gamma}(n_1,n_1,\dots,n_m,n_m)|$ denote the number of complete diagrams in $\bar{\Gamma}(n_1,n_1,\dots,n_m,n_m)$, and put

$$C(n_1,\dots,n_m)=\frac{|\bar{\Gamma}(n_1,n_1,\dots,n_m,n_m)|}{n_1!\cdots n_m!}.$$

In the special case $n_1=\cdots=n_m=n$ let $\bar{C}(n,m)=C(n_1,\dots,n_m)$. Then

Corollary 5.6.

$$E\left[(n_1!I_G(h_1))^2\cdots(n_m!I_G(h_m))^2\right]$$
$$\le C(n_1,\dots,n_m)E(n_1!I_G(h_1))^2\cdots(n_m!E(I_G(h_m))^2.$$

In particular,

$$E\left[(n!I_G(h))^{2m}\right]\le \bar C(n,m)(E(n!I_G(h))^2)^m \quad \text{if } h\in\mathscr{H}_G^n.$$

Corollary 5.6 follows immediately from Corollary 5.4 by applying it first for the sequence $h_1,h_1,\dots,h_m,h_m$ and then for the pair h_j,h_j which yields that

$$E(n_j!I_G(h_j))^2=n_j!\|h_j\|^2,\quad 1\le j\le m.$$

One only has to observe that $|h_\gamma|\le\|h_1\|^2\cdots\|h_m\|^2$ for all complete diagrams by Part (A) of Theorem 5.3.

The inequality in Corollary 5.6 is sharp. If G is a finite measure and $h_1\in H_G^{n_1},\dots,\ h_m\in H_G^{n_m}$ are constant functions, then equality can be written in Corollary 5.6. We remark that in this case $I_G(h_1),\dots,I_G(h_m)$ are constant times the n_1-th,..., n_m-th Hermite polynomials of the same standard normal random variable. Let us emphasize that the constant $C(n_1,\dots,n_m)$ depends only on the parameters $n_1,\dots,n_m$ and not on the form of the functions $h_1,\dots,h_m$. The function $C(n_1,\dots,n_m)$ is monotone in its arguments. The following argument shows that

$$C(n_1+1,n_2,\dots,n_m)\ge C(n_1,\dots,n_m)$$

Let us call two complete diagrams in $\bar\Gamma(n_1,n_1,\dots,n_m,n_m)$ or in $\bar\Gamma(n_1+1,n_1+1,\dots,n_m,n_m)$ equivalent if they can be transformed into each other by permuting the vertices $(1,1),\dots,(1,n_1)$ in $\bar\Gamma(n_1,n_1,\dots,n_m,n_m)$ or the vertices $(1,1),\dots,(1,n_1+1)$ in $\bar\Gamma(n_1+1,n_1+1,\dots,n_m,n_m)$. The equivalence classes have $n_1!$ elements in the first case and $(n_1+1)!$ elements in the second one. Moreover, the number of equivalence classes is less in the first case than in the second one. (They would agree if we counted only those equivalence classes in the second case which contain a diagram where $(1,n_1+1)$ and $(2,n_1,1)$ are connected by an edge. Hence

$$\frac{1}{n_1!}|\bar\Gamma(n_1,n_1,\dots,n_m,n_m)|\le\frac{1}{(n_1+1)!}|\bar\Gamma(n_1+1,n_1+1,\dots,n_m,n_m)|$$

as we claimed.

The next result may better illuminate the content of Corollary 5.6.

Corollary 5.7. *Let $\xi_1,\dots,\xi_k$ be a normal random vector, and $P(x_1,\dots,x_k)$ a polynomial of degree n. Then*

$$E\left[P(\xi_1,\dots,\xi_k)^{2m}\right] \le \bar{C}(n,m)(n+1)^m \left(EP(\xi_1,\dots,\xi_k)^2\right)^m$$

with the constant $\bar{C}(n,m)$ *introduced before Corollary 5.6.*

The multiplying constant $\bar{C}(n,m)(n+1)^m$ is not sharp in this case.

Proof of Corollary 5.7. We can write $\xi_j = \int f_j(x)Z(dx)$ with some $f_j \in \mathcal{H}^1$, $j = 1,2,\dots,k$, where $Z(dx)$ is the same as in the proof of Corollary 5.5. There exist some $h_j \in \mathcal{H}^j$, $j = 0,1,\dots,n$, such that

$$P(\xi_1,\dots,\xi_k) = \sum_{j=0}^{n} j!I(h_j).$$

Then

$$EP(\xi_1,\dots,\xi_k)^{2m} = E\left[\left(\sum_{j=0}^{n} j!I(h_j)\right)^{2m}\right] \le (n+1)^m E\left[\sum_{j=0}^{n}(j!I(h_j))^2\right]^m$$

$$\le (n+1)^m \sum_{p_1+\dots+p_n=m} C(p_1,\dots,p_n)(EI(h_0)^2)^{p_0}\cdots(En!I(h_n)^2)^{p_n}\frac{m!}{p_1!\cdots p_n!}$$

$$\le (n+1)^m\bar{C}(n,m) \sum_{p_1+\dots+p_n=m} (EI(h_0)^2)^{p_0}\cdots(EI(n!h_n)^2)^{p_n}\frac{m!}{p_1!\cdots p_n!}$$

$$= (n+1)^m\bar{C}(n,m)\left[\sum E(j!I(h_j))^2\right]^m$$

$$= (n+1)^m\bar{C}(n,m)\left(EP(\xi_1,\dots,\xi_k)^2\right)^m. \qquad \square$$

Chapter 6
Subordinated Random Fields: Construction of Self-similar Fields

Let X_n, $n \in \mathbb{Z}_\nu$, be a discrete stationary Gaussian random field with a non-atomic spectral measure, and let the random field ξ_n, $n \in \mathbb{Z}_\nu$, be subordinated to it. Let Z_G denote the random spectral measure adapted to the random field X_n. By Theorem 4.2 the random variable ξ_0 can be represented as

$$\xi_0 = f_0 + \sum_{k=1}^{\infty} \frac{1}{k!} \int f_k(x_1, \dots, x_k) Z_G(dx_1) \dots Z_G(dx_k)$$

with an appropriate $f = (f_0, f_1, \dots) \in \operatorname{Exp} \mathcal{H}_G$ in a unique way. This formula together with Theorem 4.4 yields the following

Theorem 6.1. *A random field ξ_n, $n \in \mathbb{Z}_\nu$, subordinated to the stationary Gaussian random field X_n, $n \in \mathbb{Z}_\nu$, with non-atomic spectral measure can be written in the form*

$$\xi_n = f_0 + \sum_{k=1}^{\infty} \frac{1}{k!} \int e^{i((n,x_1+\dots+x_k)} f_k(x_1, \dots, x_k) Z_G(dx_1) \dots Z_G(dx_k), \quad n \in Z_\nu, \tag{6.1}$$

with some $f = (f_0, f_1, \dots) \in \operatorname{Exp} \mathcal{H}_G$, where Z_G is the random spectral measure adapted to the random field X_n. This representation is unique. It is also clear that formula (6.1) defines a subordinated field for all $f \in \operatorname{Exp} \mathcal{H}_G$.

Let G denote the spectral measure of the underlying stationary Gaussian random field. If it has the property $G(\{x\colon\ x_p = u\}) = 0$ for all $u \in R^1$ and $1 \le p \le \nu$, where $x = (x_1, \dots, x_\nu)$ (this is a strengthened form of the non-atomic property of G), then the functions

$$\bar{f}_k(x_1, \dots, x_k) = f_k(x_1, \dots, x_k) \tilde{\chi}_0^{-1}(x_1 + \dots + x_k), \quad k = 1, 2, \dots,$$

P. Major, *Multiple Wiener-Itô Integrals*, Lecture Notes in Mathematics 849, DOI 10.1007/978-3-319-02642-8_6,

are meaningful, as functions in the measure space $(R^{k\nu}, \mathscr{B}^{k\nu}, G^k)$, where $\tilde{\chi}_n(x) = e^{i(n,x)} \prod_{p=0}^{\nu} \frac{e^{ix^{(p)}}-1}{ix^{(p)}}$, $n \in \mathbb{Z}_\nu$, denotes the Fourier transform of the indicator function of the ν-dimensional unit cube $\prod_{p=1}^{\nu} [n^{(p)}, n^{(p)} + 1]$. Then the random variable ξ_n in formula (6.1) can be rewritten in the form

$$\xi_n = f_0 + \sum_{k=1}^{\infty} \frac{1}{k!} \int \tilde{\chi}_n(x_1 + \cdots + x_k) \bar{f}_k(x_1, \dots, x_k) Z_G(dx_1) \dots Z_G(dx_k), \quad n \in Z_\nu. \tag{6.2}$$

Hence the following Theorem 6.1′ can be considered as the continuous time version of Theorem 6.1.

Theorem 6.1′. *Let a generalized random field $\xi(\varphi)$, $\varphi \in \mathscr{S}$, be subordinated to a stationary Gaussian generalized random field $X(\varphi)$, $\varphi \in \mathscr{S}$. Let G denote the spectral measure of the field $X(\varphi)$, and let Z_G be the random spectral measure adapted to it. Let the spectral measure G be non-atomic. Then $\xi(\varphi)$ can be written in the form*

$$\xi(\varphi) = f_0 \cdot \tilde{\varphi}(0) + \sum_{k=1}^{\infty} \frac{1}{k!} \int \tilde{\varphi}(x_1 + \cdots + x_k) f_k(x_1, \dots, x_k) Z_G(dx_1) \dots Z_G(dx_k), \tag{6.3}$$

where the functions f_k are invariant under all permutations of their variables,

$$f_k(-x_1, \dots, -x_k) = \overline{f_k(x_1, \dots, x_k)}, \quad k = 1, 2, \dots,$$

and

$$\sum_{k=1}^{\infty} \frac{1}{k!} \int (1+|x_1+\cdots+x_k|^2)^{-p} |f_k(x_1+\cdots+x_k)|^2 G(dx_1) \dots G(dx_k) < \infty \tag{6.4}$$

with an appropriate number $p > 0$. This representation is unique.

Contrariwise, all random fields $\xi(\varphi)$, $\varphi \in \mathscr{S}$, defined by formulas (6.3) and (6.4) are subordinated to the stationary, Gaussian random field $X(\varphi)$, $\varphi \in \mathscr{S}$.

Proof of Theorem 6.1′. The proof is based on the same ideas as the proof of Theorem 6.1. But here we also adapt some arguments from the theory of generalized functions (see [16]). In particular, we exploit the following continuity property of generalized random fields and subordinated generalized random fields. If $\varphi_n \to \varphi$ in the topology of the Schwartz space $\mathscr{S}$, and $X(\varphi)$, $\varphi \in \mathscr{S}$, is a generalized random field, then $X(\varphi_n) \Rightarrow X(\varphi)$ stochastically. If $X(\varphi)$, $\varphi \in \mathscr{S}$, is a generalized Gaussian random field, then also the relation $E[X(\varphi_n) - X(\varphi)]^2 \to 0$ holds in this case.

Similarly, if $\xi(\varphi)$, $\varphi \in \mathcal{S}$, is a subordinated generalized random field, and $\varphi_n \to \varphi$, then $E[\xi(\varphi_n) - \xi(\varphi)]^2 \to 0$ by the definition of subordinated fields.

It can be seen with some work that a random field $\xi(\varphi)$, $\varphi \in \mathcal{S}$, defined by (6.3) and (6.4) is subordinated to $X(\varphi)$. One has to check that the definition of $\xi(\varphi)$ in formula (6.3) is meaningful for all $\varphi \in \mathcal{S}$, because of (6.4), $\xi(T_t\varphi) = T_t\xi(\varphi)$ for all shifts T_t, $t \in R^\nu$, by Theorem 4.4, and also the following continuity property holds. For all $\varepsilon > 0$ there is a small neighbourhood H of the origin in the space $\mathcal{S}$ such that if $\varphi = \varphi_1 - \varphi_2 \in H$ for some $\varphi_1, \varphi_2 \in \mathcal{S}$ then $E[\xi(\varphi_1) - \xi(\varphi_2)]^2 = E\xi(\varphi)^2 < \varepsilon^2$.

Since the Fourier transform $\varphi(\cdot) \to \tilde{\varphi}(\cdot)$ is a bicontinuous map in $\mathcal{S}$, to prove the above continuity property it is enough to check that $E\xi(\varphi)^2 < \varepsilon^2$ if $\tilde{\varphi} \in H$ for an appropriate small neighbourhood H of the origin in $\mathcal{S}$. But this relation holds with the choice $H = \{\varphi\colon (1+|x|^2)^p|\varphi(x)| \le \frac{\varepsilon^2}{K} \text{ for all } x \in R^\nu\}$ with a sufficiently large $K > 0$ because of condition (6.4).

To prove that all subordinated fields have the above representation observe that the relation

$$\xi(\varphi) = \Psi_{\varphi,0} + \sum_{k=1}^{\infty} \frac{1}{k!} \int \Psi_{\varphi,k}(x_1,\dots,x_k) Z_G(dx_1)\dots Z_G(dx_k) \tag{6.5}$$

holds for all $\varphi \in \mathcal{S}$ with some $(\Psi_{\varphi,0}, \Psi_{\varphi,1},\dots) \in \operatorname{Exp}\mathcal{H}_G$ depending on the function φ. We are going to show that these functions $\Psi_{\varphi,k}$ can be given in the form

$$\Psi_{\varphi,k}(x_1,\dots,x_k) = f_k(x_1,\dots,x_k)\cdot\tilde{\varphi}(x_1+\dots+x_k), \quad k=1,2,\dots,$$

with some functions $f_k \in \mathcal{B}^{k\nu}$, and

$$\Psi_{\varphi,0} = f_0 \cdot \tilde{\varphi}(0)$$

for all $\varphi \in \mathcal{S}$ with a sequence of functions $f_0, f_1, \dots$ not depending on φ.

To show this let us choose a $\varphi_0 \in \mathcal{S}$ such that $\tilde{\varphi}_0(x) > 0$ for all $x \in R^\nu$. (We can make for instance the choice $\varphi_0(x) = e^{-(x,x)}$.) We claim that the finite linear combinations $\sum a_p\varphi_0(x - t_p) = \sum a_p T_{t_p}\varphi_0(x)$ are dense in $\mathcal{S}$. To prove this it is enough to show that if the Fourier transform $\tilde{\psi}$ of a function $\psi \in \mathcal{S}$ has a compact support, then in every open neighbourhood of ψ (in the topology of the space $\mathcal{S}$) there is a function of the form $\sum a_p\varphi_0(x - t_p)$. Indeed, this implies that the above introduced linear combinations constitute a dense subclass of $\mathcal{S}$, since the functions ψ with the above property are dense in $\mathcal{S}$. (The statement that these functions ψ are dense in $\mathcal{S}$ is equivalent to the statement that their Fourier transforms $\tilde{\psi}$ are dense in the space $\tilde{\mathcal{S}} \subset \mathcal{S}^c$ consisting of the Fourier transforms of the (real valued) functions in the space $\mathcal{S}$.) We have $\frac{\tilde{\psi}}{\tilde{\varphi}_0} \in \mathcal{S}^c$ for such functions ψ, where $\mathcal{S}^c$ denotes the Schwartz-space of complex valued, at infinity strongly decreasing, smooth functions again, because $\tilde{\varphi}_0(x) \ne 0$, and $\tilde{\psi}$ has a compact support. There exists a function $\chi \in \mathcal{S}$ such that $\tilde{\chi} = \frac{\tilde{\psi}}{\tilde{\varphi}_0}$. (Here we exploit that the space of Fourier transforms of the functions from $\mathcal{S}$ agrees with the space of those functions $f \in \mathcal{S}^c$ for which $f(-x) = \overline{f(x)}$.) Therefore $\psi(x) = \chi * \varphi_0(x) = \int \chi(t)\varphi_0(x-t)\,dt$, where $*$ denotes convolution. It can be seen by exploiting this relation together with the rapid decrease of χ and φ_0 together of its derivatives at infinity, and approximating the integral defining the convolution by an appropriate finite sum that for all integers $r > 0$, $s > 0$ and real numbers $\varepsilon > 0$ there exists a finite linear combination $\hat{\psi}(x) = \hat{\psi}_{r,s,\varepsilon}(x) = \sum\limits_p a_p\varphi_0(x - t_p)$ such that $(1 + |x|^s)|\psi(x) - \hat{\psi}(x)| < \varepsilon$ for all $x \in R^\nu$, and the same estimate holds for all derivatives of $\psi(x) - \hat{\psi}(x)$ of order less than r.

I only briefly explain why such an approximation exists. Some calculation enables us to reduce this statement to the case when $\psi = \chi * \varphi_0$ with a function $\chi \in \mathcal{D}$, which has compact support. To give the desired approximation choose a small number $\delta > 0$, introduce the cube $\Delta = \Delta(\delta) = [-\delta, \delta)^\nu \subset R^\nu$ and define the vectors $k(\delta) = (2k_1\delta, \dots, 2k_\nu\delta) \in R^\nu$ for all $k = (k_1, \dots, k_\nu) \in \mathbb{Z}_\nu$. Given a fixed vector $x \in R^\nu$ let us define the vector $u(x) \in R^\nu$ for all $u \in R^\nu$ as $u(x) = x + k(\delta)$ with that vector $k \in \mathbb{Z}_\nu$ for which $x + k(\delta) - u \in \Delta$, and put $\varphi_{0,x}(u) = \varphi_0(u(x))$. It can be seen that $\hat{\psi}(x) = \chi * \varphi_{0,x}(x)$ is a finite linear combination of numbers of the form $\varphi_0(x - t_k)$ (with $t_k = k(\delta)$) with coefficients not depending on x. Moreover, if $\delta > 0$ is chosen sufficiently small (depending on r, s and ε), then $\hat{\psi}(x) = \hat{\psi}_{r,s,\varepsilon}(x)$ has all properties we demanded.

The above argument implies that there is a sequence of functions $\hat{\psi}_{r,s,\varepsilon}$ which converges to the function ψ in the topology of the space $\mathcal{S}$. As a consequence, the finite linear combinations $\sum a_p\varphi_0(x - t_p)$ are dense in $\mathcal{S}$.

Define

$$f_k(x_1, \dots, x_k) = \frac{\Psi_{\varphi_0,k}(x_1, \dots, x_k)}{\tilde{\varphi}_0(x_1 + \dots + x_k)}, \quad k = 1, 2, \dots, \quad \text{and } f_0 = \frac{\Psi_{\varphi_0,0}}{\tilde{\varphi}_0(0)}.$$

If $\varphi(x) = \sum a_p\varphi_0(x - t_p) = \sum a_p T_{t_p}\varphi_0(x)$, and the sum defining φ is finite, then by Theorem 4.4

$$\begin{aligned}\xi(\varphi) &= \left(\sum a_p\right) f_0 \cdot \tilde{\varphi}_0(0) + \sum_{k=1}^{\infty} \frac{1}{k!} \int \sum_p a_p e^{i(t_p, x_1 + \dots + x_k)} \tilde{\varphi}_0(x_1 + \dots + x_k) \\ &\qquad \cdot f_k(x_1, \dots, x_k) Z_G(dx_1) \dots Z_G(dx_k) \\ &= f_0 \cdot \tilde{\varphi}(0) + \sum_{k=1}^{\infty} \frac{1}{k!} \int \tilde{\varphi}(x_1 + \dots + x_k) f_k(x_1, \dots, x_k) Z_G(dx_1) \dots Z_G(dx_k).\end{aligned}$$

Relation (6.5) holds for all $\varphi \in \mathcal{S}$, and there exists a sequence of functions $\varphi_j(x) = \sum\limits_p a_p^{(j)}\varphi_0(x - t_p^{(j)}) \in \mathcal{S}$ satisfying (6.3) such that $\varphi_j \to \varphi$ in the topology of $\mathcal{S}$. This implies that $\lim E[\xi(\varphi_j) - \xi(\varphi)]^2 \to 0$, and in particular $EI_G(\Psi_{\varphi,k} - \hat{\varphi}_{j,k} f_k)^2 \to 0$ with $\hat{\varphi}_{j,k}(x_1, \dots, x_k) = \tilde{\varphi}_j(x_1 + \dots + x_k)$ as $j \to \infty$ for all $k = 1, 2, \dots$. In the subsequent steps of the proof we restrict the domain of integration to bounded sets A, because this enables us to carry out some limiting procedures needed in our argument. We can write that

$$\int_A |\Psi_{\varphi,k}(x_1,\dots,x_k) - \tilde{\varphi}_j(x_1+\cdots+x_k) f_k(x_1,\dots,x_k)|^2 G(dx_1)\dots G(dx_k) \to 0$$

as $j \to \infty$ for all k and for all bounded sets $A \in R^{k\nu}$. On the other hand,

$$\int_A |\tilde{\varphi}(x_1+\cdots+x_k) - \tilde{\varphi}_j(x_1+\cdots+x_k)|^2 |f_k(x_1,\cdots,x_k)|^2 G(dx_1)\dots G(dx_k) \to 0,$$

since $\tilde{\varphi}_j(x) - \tilde{\varphi}(x) \to 0$ in the supremum norm if $\tilde{\varphi}_j \to \tilde{\varphi}$ in the topology of $\mathcal{S}$, and the property $\tilde{\varphi}_0(x) > 0$ (of the function $\tilde{\varphi}_0$ appearing in the definition of the function f_k) together with the continuity of $\tilde{\varphi}_0$ and the inequality $EI_G(\hat{\varphi}_{0,k} f_k)^2 < \infty$ imply that $\int_A |f_k(x_1,\dots,x_k)|^2 G(dx_1)\dots G(dx_k) < \infty$ on all bounded sets A. The last two relations yield that

$$\Psi_{\varphi,k}(x_1,\dots,x_k) = \tilde{\varphi}(x_1+\cdots+x_k) f_k(x_1,\dots,x_k), \quad k = 1,2,\dots,$$

since both sides of this identity is the limit of the sequence

$$\tilde{\varphi}_j(x_1+\cdots+x_k) f_k(x_1,\dots,x_k), \quad j = 1,2,\dots$$

in the $L^2_{G^k_A}$ norm, where G^k_A denotes the restriction of the measure G^k to the set A.

Similarly,

$$\psi_{\varphi,0} = \tilde{\varphi}(0) f_0.$$

These relations imply (6.3).

To complete the proof of Theorem 6.1′ we show that (6.4) follows from the continuity of the transformation $F\colon\ \varphi \to \xi(\varphi)$ from the space $\mathcal{S}$ into the space $L_2(\Omega, \mathcal{A}, P)$.

We recall that the transformation $\varphi \to \tilde{\varphi}$ is bicontinuous in $\mathcal{S}^c$. Hence for a subordinated field $\xi(\varphi)$, $\varphi \in \mathcal{S}$, the transformation $\tilde{\varphi} \to \xi(\varphi)$ is a continuous map from the space of the Fourier transforms of the functions in the space $\mathcal{S}$ to $L_2(\Omega, \mathcal{A}, P)$. This continuity implies that there exist some integers $p > 0$, $r > 0$ and real number $\delta > 0$ such that if

$$(1+|x^2|)^p \left| \frac{\partial^{s_1+\cdots+s_\nu}}{\partial x^{(1)^{s_1}} \dots \partial x^{(\nu)^{s_\nu}}} \tilde{\varphi}(x) \right| < \delta \quad \text{for all } s_1+\cdots+s_\nu \le r, \tag{6.6}$$

then $E\xi(\varphi)^2 \le 1$.

Let us choose a function $\psi \in \mathcal{S}$ such that ψ has a compact support, $\psi(x) = \psi(-x)$, $\psi(x) \ge 0$ for all $x \in R^\nu$, and $\psi(x) = 1$ if $|x| \le 1$. (There exist such functions.) Define the functions $\tilde{\varphi}_m(x) = C(1+|x|^2)^{-p}\psi(\frac{x}{m})$. Then $\varphi_m \in \mathcal{S}$, since its Fourier transform $\tilde{\varphi}_m$ is an even function, and it is in the space $\mathcal{S}$ being an infinite many times differentiable function with compact support. Moreover, φ_m satisfies (6.6) for all $m = 1, 2, \dots$ if the number $C > 0$ in its definition is chosen sufficiently small. This number C can be chosen independently of m. (To see this observe that $(1 + |x^2|)^{-p}$ together with all of its derivatives of order not bigger than r can be bounded by $\frac{C(p,r)}{(1+|x|^2)^p}$ with an appropriate constant $C(p,r)$.) Hence

$$E\xi(\varphi_m)^2 = \sum \frac{1}{k!} \int |\tilde{\varphi}_m(x_1 + \cdots + x_k)|^2 |f_k(x_1, \cdots, x_k)|^2 G(dx_1) \dots G(dx_k) \le 1$$

for all $m = 1, 2, \dots$.

As $\tilde{\varphi}_m(x) \to C(|1 + |x|^2)^{-p}$ as $m \to \infty$, and $\tilde{\varphi}_k(x) \ge 0$, an $m \to \infty$ limiting procedure in the last relation together with Fatou's lemma imply that

$$C \sum \frac{1}{k!} \int (1 + |x_1 + \cdots + x_k)|^2)^{-p} |f_k(x_1, \cdots, x_k)|^2 G(dx_1) \dots G(dx_k) \le 1.$$

Theorem 6.1′ is proved. □

We shall call the representations given in Theorems 6.1 and 6.1′ the canonical representation of a subordinated field. From now on we restrict ourselves to the case $E\xi_n = 0$ or $E\xi(\varphi) = 0$ respectively, i.e. to the case when $f_0 = 0$ in the canonical representation. If

$$\xi(\varphi) = \sum_{k=1}^{\infty} \frac{1}{k!} \int \tilde{\varphi}(x_1 + \cdots + x_k) f_k(x_1, \dots, x_k) Z_G(dx_1) \dots Z_G(dx_k),$$

then

$$\xi(\varphi_t^A) = \sum_{k=1}^{\infty} \frac{1}{k!} \frac{t^\nu}{A(t)} \int \tilde{\varphi}(t(x_1 + \cdots + x_k)) f_k(x_1, \dots, x_k) Z_G(dx_1) \dots Z_G(dx_k)$$

with the function φ_t^A defined in (1.3). Define the spectral measures G_t by the formula $G_t(A) = G(tA)$. Then we have by Lemma 4.6

$$\xi(\varphi_t^A) \stackrel{\Delta}{=} \sum_{k=1}^{\infty} \frac{1}{k!} \frac{t^\nu}{A(t)} \int \tilde{\varphi}(x_1 + \cdots + x_k) f_k\left(\frac{x_1}{t}, \dots, \frac{x_k}{t}\right) Z_{G_t}(dx_1) \dots Z_{G_t}(dx_k).$$

If $G(tB) = t^{2\kappa} G(B)$ with some $\kappa > 0$ for all $t > 0$ and $B \in \mathcal{B}^\nu$, $f_k(\lambda x_1, \dots, \lambda x_k) = \lambda^{\nu - \kappa k - \alpha} f_k(x_1, \dots, x_k)$, and $A(t)$ is chosen as $A(t) = t^\alpha$, then Theorem 4.5 (with the choice $G'(B) = G(tB) = t^{2\kappa} G(B)$) implies that $\xi(\varphi_t^A) \stackrel{\Delta}{=} \xi(\varphi)$. Hence we obtain the following

Theorem 6.2. *Let a generalized random field $\xi(\varphi)$ be given by the formula*

$$\xi(\varphi) = \sum_{k=1}^{\infty} \frac{1}{k!} \int \tilde{\varphi}(x_1 + \cdots + x_k) f_k(x_1, \dots, x_k) Z_G(dx_1) \dots Z_G(dx_k). \quad (6.7)$$

If $f_k(\lambda x_1, \dots, \lambda x_k) = \lambda^{\nu - \kappa k - \alpha} f_k(x_1, \dots, x_k)$ for all k, $(x_1, \dots, x_k) \in R^{k\nu}$ and $\lambda > 0$, $G(\lambda A) = \lambda^{2\kappa} G(A)$ for all $\lambda > 0$ and $A \in \mathcal{B}^\nu$, then ξ is a self-similar random field with parameter α.

The discrete time version of this result can be proved in the same way. It states the following

Theorem 6.2′. *If a discrete random field ξ_n, $n \in \mathbb{Z}_\nu$, has the form*

$$\xi_n = \sum_{k=1}^{\infty} \frac{1}{k!} \int \tilde{\chi}_n(x_1 + \cdots + x_k) f_k(x_1, \dots, x_k) Z_G(dx_1) \dots Z_G(dx_k), \quad n \in \mathbb{Z}_\nu, \quad (6.8)$$

and $f_k(\lambda x_1, \dots, \lambda x_k) = \lambda^{\nu - \kappa k - \alpha} f_k(x_1, \dots, x_k)$ for all k, $G(\lambda A) = \lambda^{2\kappa} G(A)$, then ξ_n is a self-similar random field with parameter α.

Theorems 6.2 and 6.2′ enable us to construct self-similar random fields. Nevertheless, we have to check whether formulas (6.7) and (6.8) are meaningful. The hard part of this problem is to check whether

$$\sum \frac{1}{k!} \int |\tilde{\chi}_n(x_1 + \cdots + x_k)|^2 |f_k(x_1, \dots, x_k)|^2 G(dx_1) \dots G(dx_k) < \infty$$

in the discrete time case or whether

$$\sum \frac{1}{k!} \int |\tilde{\varphi}(x_1 + \cdots + x_k)|^2 |f_k(x_1, \dots, x_k)|^2 G(dx_1) \dots G(dx_k) < \infty \quad \text{for all } \varphi \in \mathcal{S}$$

in the generalized field case.

It is rather hard to decide in the general case when these expressions are finite. The next result enables us to prove the finiteness of these expressions in some interesting cases.

Let us define the measure G

$$G(A) = \int_A |x|^{2\kappa - \nu} a\left(\frac{x}{|x|}\right) dx, \quad A \in \mathcal{B}^\nu, \quad (6.9)$$

where $a(\cdot)$ is a non-negative, measurable and even function on the ν-dimensional unit sphere $S_{\nu-1}$, and $\kappa > 0$. (The condition $\kappa > 0$ is imposed to guarantee the relation $G(A) < \infty$ for all bounded sets $A \in \mathcal{B}^\nu$.) We prove the following

Proposition 6.3. *Let the measure G be defined in formula (6.9).*

(a) *If the function* $a(\cdot)$ *is bounded on the unit sphere* $S_{\nu-1}$, *and* $\frac{\nu}{k} > 2\kappa > 0$, *then*

$$D(n) = \int |\tilde{\chi}_n(x_1 + \cdots + x_k)|^2 G(dx_1) \dots G(dx_k) < \infty \quad \textit{for all } n \in \mathbb{Z}_\nu,$$

and

$$\begin{aligned} D(\varphi) &= \int |\tilde{\varphi}(x_1 + \cdots + x_k)|^2 G(dx_1) \dots G(dx_k) \\ &\leq C \int (1 + |x_1 + \cdots + x_k|^2)^{-p} G(dx_1) \dots G(dx_k) < \infty \end{aligned}$$

for all $\varphi \in \mathcal{S}$ *and* $p > \frac{\nu}{2}$ *with some* $C = C(\varphi, p) < \infty$.

(b) *If there is a constant* $C > 0$ *such that* $a(x) > C$ *in a neighbourhood of a point* $x_0 \in S_{\nu-1}$, *and either* $2\kappa \leq 0$ *or* $2\kappa \geq \frac{\nu}{k}$, *then the integrals* $D(n)$ *are divergent, and the same relation holds for* $D(\varphi)$ *with some* $\varphi \in \mathcal{S}$.

Proof of Proposition 6.3. Proof of Part (a).
We may assume that $a(x) = 1$ for all $x \in S_{\nu-1}$. Define

$$J_{\kappa,k}(x) = \int_{x_1 + \cdots + x_k = x} |x_1|^{2\kappa-\nu} \cdots |x_k|^{2\kappa-\nu} \, dx_1 \dots dx_k, \qquad x \in R^\nu,$$

for $k \geq 2$, where $dx_1 \dots dx_k$ denotes the Lebesgue measure on the hyperplane $x_1 + \cdots + x_k = x$, and let $J_{\kappa,1}(x) = |x|^{2\kappa-\nu}$. We have

$$J_{\kappa,k}(\lambda x) = |\lambda|^{k(2\kappa-\nu)+(k-1)\nu} J_{\kappa,k}(x), = |\lambda|^{2k\kappa-\nu} J_{\kappa,k}(x), \qquad x \in R^\nu \ \lambda > 0,$$

because of the homogeneity of the integral. We can write, because of (6.9) with $a(x) \equiv 1$

$$D(n) = \int_{R^\nu} |\tilde{\chi}_n(x)|^2 J_{\kappa,k}(x) \, dx, \tag{6.10}$$

and

$$D(\varphi) = \int_{R^\nu} |\tilde{\varphi}(x)|^2 J_{\kappa,k}(x) \, dx.$$

We prove by induction on k that

$$J_{\kappa,k}(x) \leq C(\kappa, k) |x|^{2\kappa k-\nu} \tag{6.11}$$

with an appropriate constant $C(\kappa, k) < \infty$ if $\frac{\nu}{k} > 2\kappa > 0$.

Inequality (6.11) holds for $k = 1$, and we have

$$J_{\kappa,k}(x) = \int J_{\kappa,k-1}(y)|x-y|^{2\kappa-\nu}\,dy$$

for $k \geq 2$. Hence

$$\begin{aligned}J_{\kappa,k}(x) &\leq C(\kappa,k-1)\int |y|^{(2\kappa(k-1)-\nu}|x-y|^{2\kappa-\nu}\,dy\\ &= C(\kappa,k-1)|x|^{2\kappa k-\nu}\int |y|^{(2\kappa(k-1)-\nu}\left|\frac{x}{|x|}-y\right|^{2\kappa-\nu}dy = C(\kappa,k)|x|^{2\kappa k-\nu},\end{aligned}$$

since $\int |y|^{(2\kappa(k-1)-\nu}\left|\frac{x}{|x|}-y\right|^{2\kappa-\nu}dy < \infty$.

The last integral is finite, since its integrand behaves at zero asymptotically as $C|y|^{2\kappa(k-1)-\nu}$, at the point $e = \frac{x}{|x|} \in S_{\nu-1}$ as $C_2|y-e|^{2\kappa-\nu}$ and at infinity as $C_3|y|^{2\kappa k-2\nu}$. Relations (6.10) and (6.11) imply that

$$\begin{aligned}D(n) &\leq C'\int |\tilde{\chi}_0(x)|^2|x|^{2\kappa k-\nu}\,dx \leq C''\int |x|^{2\kappa k-\nu}\prod_{l=1}^{\nu}\frac{1}{1+|x^{(l)}|^2}\,dx\\ &\leq C'''\int_{|x^{(1)}|=\max\limits_{1\leq l\leq\nu}|x^{(l}|}|x^{(1)}|^{2\kappa k-\nu}\prod_{l=1}^{\nu}\frac{1}{1+|x^{(l)}|^2}\,dx\\ &= \sum_{p=0}^{\infty}C'''\int_{|x^{(1)}|=\max\limits_{1\leq l\leq\nu}|x^{(l}|,\ 2^p\leq|x^{(1)}|<2^{p+1}} + C'''\int_{|x^{(1)}|=\max\limits_{1\leq l\leq\nu}|x^{(l}|,|x^{(1)}|<1}.\end{aligned}$$

The second term in the last sum can be simply bounded by a constant, since $B = \left\{x\colon |x^{(1)}| = \max\limits_{1\leq l\leq\nu}|x^{(l}|,\ |x^{(1)}|<1\right\} \subset \{x\colon\ |x| \leq \sqrt{\nu}\}$, and $|x^{(1)}|^{2\kappa k-\nu}\prod\limits_{l=1}^{\nu}\frac{1}{1+|x^{(l)}|^2} \leq \text{const.}\,|x|^{2\kappa k-\nu}$ on the set B. Hence

$$D(n) \leq C_1\sum_{p=0}^{\infty}2^{p(2\kappa k-\nu)}\left[\int_{-\infty}^{\infty}\frac{1}{1+x^2}\,dx\right]^{\nu} + C_2 < \infty.$$

We have $|\varphi(x)| \leq C(1+|x^2|)^{-p}$ with some $C>0$ and $D>0$ if $\varphi \in \mathscr{S}$. The proof of the estimate $D(\varphi)<\infty$ for $\varphi \in \mathscr{S}$ is similar but simpler.

Proof of Part (b). Define, similarly to the function $J_{\kappa,k}$ the function

$$J_{\kappa,k,a}(x) = \int_{x_1+\cdots+x_k=x}|x_1|^{2\kappa-\nu}a\left(\frac{x_1}{|x_1|}\right)\cdots|x_k|^{2\kappa-\nu}a\left(\frac{x_k}{|x_k|}\right)dx_1\ldots dx_k,\quad x\in R^{\nu},$$

for $k \geq 2$, where $dx_1 \dots dx_k$ denotes the Lebesgue measure on the hyperplane $x_1 + \cdots + x_k = x$, and put $J_{1,a}(x) = x^{2\kappa-\nu} a(\frac{x}{|x|})$. We can prove by induction with respect to k that

$$J_{\kappa,k,a}(x) \geq \int_{y:\, (\frac{1}{2}-\alpha)|x|<|y|<(\frac{1}{2}+\alpha)|x|} J_{\kappa,k-1,a}(y) a\left(\frac{x-y}{|x-y|}\right) |x-y|^{2\kappa-\nu}\, dy$$

$$\geq C(\kappa,k,a(\cdot)) \int_{y:\, (\frac{1}{2}-\alpha)|x|<|y|<(\frac{1}{2}+\alpha)|x|} |x|^{2\kappa(k-1)-\nu} |x-y|^{2\kappa-\nu}\, dy$$

with the choice of some number $0 < \alpha < \frac{1}{2}$ and

$$J_{\kappa,k,a}(x) \geq \int_{y:\, |y|>2x} J_{\kappa,k-1,a}(y) a\left(\frac{x-y}{|x-y|}\right) |x-y|^{2\kappa-\nu}\, dy$$

$$\geq C(\kappa,k,a(\cdot)) \int_{y:\, |y|>2x} |x-y|^{2\kappa k-2\nu}\, dy$$

with some coefficient $C(\kappa,k,a(\cdot))$ if $\frac{x}{|x|}$ is close to such a point $x_0 \in S_{\nu-1}$ in whose small neighbourhood the function $a(\cdot)$ is separated from zero. Hence by an argument similar to the one in Part (a)

$$J_{\kappa,k,a}(x) \begin{cases} \geq \bar{C}(\kappa,k)|x|^{2\kappa k-\nu} & \text{if } \frac{\nu}{k} > 2\kappa > 0, \\ = \infty & \text{if } \kappa \leq 0 \text{ or } 2\kappa \geq \frac{\nu}{k} \end{cases}$$

for such vectors $x \in R^\nu$.

Since $|\tilde{\chi}_n(x)|^2 > 0$ for almost all $x \in R^\nu$,

$$D(n) = \int |\tilde{\chi}_n(x)|^2 J_{\kappa,k,a}(x)\, dx = \infty$$

under the conditions of Part (b). Similarly $D(\varphi) = \infty$ if $|\tilde{\varphi}(x)|^2 > 0$ for almost all $x \in R^\nu$. We remark that the conditions in Part (b) can be weakened. It would have been enough to assume that $a(x) > 0$ on a set of positive Lebesgue measure in $S_{\nu-1}$. □

Theorems 6.2 and 6.2′ together with Proposition 6.3 have the following

Corollary 6.4. *The formulae*

$$\xi_n = \sum_{k=1}^{M} \int \tilde{\chi}_n(x_1 + \cdots + x_k) \prod_{l=1}^{k} \left(|x_l|^{-\kappa+(\nu-\alpha)/k} \cdot b_k\left(\frac{x_l}{|x_l|}\right)\right)$$
$$Z_G(dx_1) \dots Z_G(dx_k), \qquad n \in \mathbb{Z}_\nu,$$

and

$$\xi(\varphi) = \sum_{k=1}^{M} \int \tilde\varphi(x_1+\cdots+x_k) \prod_{l=1}^{k} \left(|x_l|^{-\kappa+(\nu-\alpha)/k} \cdot b_k\left(\frac{x_l}{|x_l|}\right)\right)$$
$$Z_G(dx_1)\dots Z_G(dx_k), \qquad \varphi \in \mathscr{S},$$

define self-similar random fields with self-similarity parameter α if G is defined by formula (6.9), the parameter α satisfies the inequality $\frac{\nu}{2} < \alpha < \nu$, and the functions $a(\cdot)$ (in the definition of the measure $G(\cdot)$ in (6.9)) and $b_1(\cdot),\dots, b_k(\cdot)$ are bounded even functions on $S_{\nu-1}$.

The following observation may be useful in the proof of Corollary 6.4. We can replace ξ_n by another random field with the same distribution. Thus we can write, by exploiting Theorem 4.5,

$$\xi_n = \sum_{k=1}^{M} \tilde\chi_n(x_1+\cdots+x_k) Z_{G'}(dx_1)\dots Z_{G'}(dx_k), \qquad n \in \mathbb{Z}_\nu,$$

with a random spectral measure $Z_{G'}$ corresponding to the spectral measure $G'(dx) = b(\frac{x}{|x|})^2|x|^{-2\kappa+2(\nu-\alpha)/k}G(dx) = a(\frac{x}{|x|})b(\frac{x}{|x|})^2|x|^{-\nu+2(\nu-\alpha)/k}\,dx$. In the case of generalized random fields a similar argument can be applied.

Remark 6.5. The estimate on $J_{\kappa,k}$ and the end of the of Part (a) in Proposition 6.3 show that the self-similar random field

$$\xi(\varphi) = \sum_{k=1}^{M} \int \tilde\varphi(x_1+\cdots+x_k)|x_1+\cdots+x_k|^p\, u\left(\frac{x_1+\cdots+x_k}{|x_1+\cdots+x_k|}\right)$$
$$\prod_{l=1}^{k} \left(|x_l|^{-\kappa+(\nu-\alpha)/k} \cdot b_k\left(\frac{x_l}{|x_l|}\right)\right) Z_G(dx_1)\dots Z_G(dx_k), \qquad \varphi \in \mathscr{S},$$

and

$$\xi_n = \sum_{k=1}^{M} \int \tilde\chi_n(x_1+\cdots+x_k)|x_1+\cdots+x_k|^p\, u\left(\frac{x_1+\cdots+x_k}{|x_1+\cdots+x_k|}\right)$$
$$\prod_{l=1}^{k} \left(|x_l|^{-\kappa+(\nu-\alpha)/k} \cdot b_k\left(\frac{x_l}{|x_l|}\right)\right) Z_G(dx_1)\dots Z_G(dx_k), \qquad n \in \mathbb{Z}_\nu,$$

are well defined if G is defined by formula (6.9), $a(\cdot)$, $b(\cdot)$ and $u(\cdot)$ are bounded even functions on $S_{\nu-1}$, $\frac{\nu}{2} < \alpha < \nu$, and $\alpha - p < \nu$ in the generalized and $\frac{\nu-1}{2} < \alpha - p < \nu$ is the discrete random field case. The self-similarity parameter of these random fields is $\alpha - p$. We remark that in the case $p > 0$ this class of self-similar fields also contains self-similar fields with self-similarity parameter less than $\frac{\nu}{2}$.

In proving the statement of Remark 6.5 we have to check the integrability conditions needed for the existence of the Wiener–Itô integrals $\xi(\varphi)$ and ξ_n. To check them it is worth remarking that in the proof of Part (a) of Proposition 6.3 we proved the estimate $J_{\bar\kappa,k}(x) \le C(\bar\kappa,k)|x|^{2\bar\kappa k-\nu}$. We want to apply this inequality in the present case with the choice $\bar\kappa = \frac{\nu-\alpha}{k}$. Then arguing similarly to the proof of Part (a) of Proposition 6.3 we get to the problem whether the relations $\int |\tilde\chi_n(x)|^2|x|^{2p+2(\nu-\alpha)-\nu}\,dx < \infty$ and $\int |\tilde\varphi(x)|^2|x|^{2p+2(\nu-\alpha)-\nu}\,dx < \infty$ if $\varphi \in \mathcal{S}$ hold under the conditions of Remark 6.5. They can be proved by means of the argument applied at the end of the proof of Part (a) of Proposition 6.3.

The following question arises in a natural way. When do different formulas satisfying the conditions of Theorems 6.2 or 6.2′ define self-similar random fields with different distributions? In particular: Are the self-similar random fields constructed via multiple Wiener–Itô integrals of order $k \ge 2$ necessarily non-Gaussian? We cannot give a completely satisfactory answer for the above question, but our former results yield some useful information.

Let us substitute the spectral measure G by G' such that $\frac{G(dx)}{G'(dx)} = |g^2(x)|^2$, $g(-x) = \overline{g(x)}$ and the functions $|x_l|^{-\kappa+(\nu-\alpha)/k}b(\frac{x_l}{|x_l|})$ by $b(\frac{x_l}{|x_l|})g(x_l)|x_l|^{-\kappa+(\nu-\alpha)/k}$ in Corollary 6.4. By Theorem 4.4 the new field has the same distribution as the original one. On the other hand, Corollary 5.4 may helps us to decide whether two random variables have different moments, and therefore different distributions. Let us consider e.g. a moment of odd order of the random variables ξ_n or $\xi(\varphi)$ defined in Corollary 6.4. It is clear that all $h_\gamma \ge 0$. Moreover, if $b_k(x)$ does not vanish for some even number k, then there exists a $h_\gamma > 0$ in the sum expressing an odd moment of ξ_n or $\xi(\varphi)$. Hence the odd moments of ξ_n or $\xi(\varphi)$ are positive in this case. This means in particular that the self-similar random fields defined in Corollary 6.4 are non-Gaussian if b_k is non-vanishing for some even k. The next result shows that the tail behaviour of multiple Wiener–Itô integrals of different order is different.

Theorem 6.6. *Let G be a non-atomic spectral measure and Z_G a random spectral measure corresponding to G. For all $h \in \mathcal{H}_G^m$ there exist some constants $K_1 > K_2 > 0$ and $x_0 > 0$ depending on the function h such that*

$$e^{-K_1 x^{2/m}} \le P(|I_G(h)| > x) \le e^{-K_2 x^{2/m}}$$

for all $x > x_0$.

Remark. As the proof of Theorem 6.6 shows the constant K_2 in the upper bound of the above estimate can be chosen as $K_m = C_m(EI_G(h)^2)^{-1/m}$ with a constant C_m depending only on the order m of the Wiener–Itô integral of $I_G(h)$. This means that for a fixed number m the constant K_2 in the above estimate can be chosen as a constant depending only on the variance of the random variable $I_G(h)$. On the other hand, no simple characterization of the constant $K_1 > 0$ appearing in the lower bound of this estimate is known.

Proof of Theorem 6.6. (a) Proof of the Upper Estimate.

We have

$$P(|I_G(h)| > x) \le x^{2N} E(I_G(h)|^{2N}).$$

By Corollary 5.6

$$E(I_G(h)|^{2N}) \le \bar{C}(m,N)[E(I_G(h)^2)]^N \le \bar{C}(m,N)C_1^N$$

with the coefficient $\bar{C}(m,N)$ appearing in this result, and by a simple combinatorial argument we obtain that

$$\bar{C}(m,N) \le \frac{(2Nm-1)(2Nm-3)\cdots 1}{(m!)^N},$$

since the numerator on the right-hand side of this inequality equals the number of complete diagrams $|\bar{\Gamma}(\underbrace{m,\dots,m}_{2N \text{ times}})|$ if vertices from the same row can also be connected. Multiplying the inequalities

$$(2nM-2j-1)(2Nm-2j-1-2N)\cdots(2Nm-2j-1-2N(m-1)) \le (2N)^m m!,$$

$j=1,\dots,N$, we obtain that

$$\bar{C}(m,N) \le (2N)^{mN}.$$

(This inequality could be sharpened, but it is sufficient for our purpose.) Choose a sufficiently small number $\alpha > 0$, and define $N = [\alpha x^{2/m}]$, where $[\cdot]$ denotes integer part. With this choice we have

$$P(|I_G(h)| > x) \le (x^{-2}(2\alpha)^m x^2)^N C_1^N = [C_1(2\alpha)^m]^N \le e^{-K_2 x^{2/m}},$$

if α is chosen in such a way that $C_1(2\alpha)^m \le \frac{1}{e}$, $K_2 = \frac{\alpha}{2}$, and $x > x_0$ with an appropriate $x_0 > 0$.

(b) Proof of the Lower Estimate.

First we reduce this inequality to the following statement. Let $Q(x_1,\dots,x_k)$ be a homogeneous polynomial of order m (the number k is arbitrary), and $\xi = (\xi_1,\dots,\xi_k)$ a k-dimensional standard normal variable. Then

$$P(Q(\xi_1,\dots,\xi_k) > x) \ge e^{-Kx^{2/m}} \tag{6.12}$$

if $x > x_0$, where the constants $K > 0$ and $x_0 > 0$ may depend on the polynomial Q.

By the results of Chap. 4, $I_G(h)$ can be written in the form

$$I_G(h) = \sum_{j_1+\dots+j_l=m} C^{k_1,\dots,k_l}_{j_1,\dots,j_l} H_{j_1}(\xi_{k_1})\cdots H_{j_k}(\xi_{k_l}), \tag{6.13}$$

where $\xi_1, \xi_2, \dots$ are independent standard normal random variables, $C^{k_1,\dots,k_l}_{j_1,\dots,j_l}$ are appropriate coefficients, and the right-hand side of (6.13) is convergent in L_2 sense. Let us fix a sufficiently large integer k, and let us consider the conditional distribution of the right-hand side of (6.13) under the condition $\xi_{k+1} = x_{k+1}, \xi_{k+2} = x_{k+2}, \dots$, where the numbers $x_{k+1}, x_{k+2}, \dots$ are arbitrary. This conditional distribution coincides with the distribution of the random variable $Q(\xi_1, \dots, \xi_k, x_{k+1}, x_{k+2}, \dots)$ with probability 1, where the polynomial Q is obtained by substituting $\xi_{k+1} = x_{k+1}$, $\xi_{k+2} = x_{k=2}, \dots$ into the right-hand side of (6.13).
In particular,

$$Q(\xi_1, \dots, \xi_k, x_{k+1}, x_{k+2}, \dots)$$

is a random polynomial with finite second moment, and as a consequence with finite coefficients for almost all vectors $(x_{k+1}, x_{k+2}, \dots)$ with respect to the distribution of the vector $(\xi_{k+1}, \xi_{k+2}, \dots)$. It is clear that all these polynomials

$$Q(\xi_1, \dots, \xi_k, x_{k+1}, x_{k+2}, \dots)$$

are of order m if k is sufficiently large. It is sufficient to prove that

$$P(|Q(\xi_1, \dots, \xi_k, x_{k+1}, x_{k+2}, \dots)| > x) \geq e^{-Kx^{2/m}}$$

for $x > x_0$, where the constants $K > 0$ and $x_0 > 0$ may depend on the polynomial Q. Write

$$Q(\xi_1, \dots, \xi_k, x_{k+1}, x_{k+2}, \dots) = Q_1(\xi_1, \dots, \xi_k) + Q_2(\xi_1, \dots, \xi_k)$$

where Q_1 is a homogeneous polynomial of order m, and Q_2 is a polynomial of order less than m. The polynomial Q_2 can be rewritten as the sum of finitely many Wiener–Itô integrals with multiplicity less than m. Hence the already proved part of Theorem 6.6 implies that

$$P(Q_2(\xi_1, \dots, \xi_k) > x) \leq e^{-\bar{q}Kx^{2/(m-1)}}.$$

(We may assume that $m \geq 2$.) Then an application of relation (6.12) to Q_1 implies the remaining part of Theorem 6.6, thus it suffices to prove (6.12).

If $Q(\cdot)$ is a polynomial of k variables, then there exist some $\alpha > 0$ and $\beta > 0$ such that

$$\lambda\left(\left|Q\left(\frac{x_1}{|x|}, \dots, \frac{x_k}{|x|}\right)\right| > \alpha\right) > \beta,$$

where $|x|^2 = \sum_{j=1}^{k} x_j^2$, and λ denotes the Lebesgue measure on the k-dimensional unit sphere S_{k-1}. Exploiting that $|\xi|$ and $\frac{\xi}{|\xi|}$ are independent, $\frac{\xi}{|\xi|}$ is uniformly distributed on the unit sphere S_{k-1}, and $P(|\xi| > x) \geq ce^{-x^2}$ for a k-dimensional standard normal random variable, we obtain that

$$P(|Q(\xi_1, \dots, \xi_k)| > x) \geq \beta P\left(|\xi|^m > \frac{x}{\alpha}\right) > e^{-Kx^{2/m}},$$

if the constants K and x are sufficiently large. Theorem 6.6 is proved. □

Theorem 6.6 implies in particular that Wiener–Itô integrals of different multiplicity have different distributions. A bounded random variable measurable with respect to the σ-algebra generated by a stationary Gaussian field can be expressed as a sum of multiple Wiener–Itô integrals. Another consequence of Theorem 6.6 is the fact that the number of terms in this sum must be infinite.

In Theorems 6.2 and 6.2′ we have defined a large class of self-similar fields. The question arises whether this class contains self-similar fields such that the distributions of their random variables tend to one (or zero) at infinity (at minus infinity) much faster than the normal distribution functions do. This question has been unsolved by now. By Theorem 6.6 such fields, if any, must be expressed as a sum of infinitely many Wiener–Itô integrals. The above question is of much greater importance than it may seem at first instant. Some considerations suggest that in some important models of statistical physics self-similar fields with very fast decreasing tail distributions appear as limit, when the so-called renormalization group transformations are applied for the probability measure describing the state of the model at critical temperature. (The renormalization group transformations are the transformations over the distribution of stationary fields induced by formula (1.1) or (1.3), when $A_N = N^\alpha$, $A(t) = t^\alpha$ with some α.) No rigorous proof about the existence of such self-similar fields is known yet. Thus the real problem behind the above question is whether the self-similar fields interesting for statistical physics can be constructed via multiple Wiener–Itô integrals.

Chapter 7
On the Original Wiener–Itô Integral

In this chapter the definition of the original Wiener–Itô integral introduced by Itô in [19] is explained. As the arguments are very similar to those of Chaps. 4 and 5 (only the notations become simpler) most proofs will be omitted.

Let a measure space $(M, \mathcal{M}, \mu)$ with a σ-finite measure μ be given. Let μ satisfy the following continuity property: For all $\varepsilon > 0$ and $A \in \mathcal{M}$, $\mu(A) < \infty$, there exist some disjoint sets $B_j \in \mathcal{M}$, $j = 1, \dots, N$, with some integer N such that $\mu(B_j) < \varepsilon$ for all $1 \le j \le N$, and $A = \bigcup\limits_{j=1}^{N} B_j$. We introduce the following definition.

Definition of (Gaussian) Random Orthogonal Measures. *A system of random variables $Z_\mu(A)$, $A \in \mathcal{M}$, $\mu(A) < \infty$, is called a Gaussian random orthogonal measure corresponding to the measure μ if*

(i) $Z_\mu(A_1), \dots, Z_\mu(A_k)$ *are independent Gaussian random variables if the sets $A_j \in \mathcal{M}$, $\mu(A_j) < \infty$, $j = 1, \dots, k$, are disjoint.*
(ii) $EZ_\mu(A) = 0$, $EZ_\mu(A)^2 = \mu(A)$.
(iii) $Z_\mu\left(\bigcup\limits_{j=1}^{k} A_j\right) = \sum\limits_{j=1}^{k} Z_\mu(A_k)$ *with probability 1 if $A_1, \dots, A_k$ are disjoint sets.*

Remark. There is the following equivalent version for the definition of random orthogonal measures: The system of random variables system of random variables $Z_\mu(A)$, $A \in \mathcal{M}$, $\mu(A) < \infty$, is a Gaussian random orthogonal measure corresponding to the measure μ if

(i′) $Z_\mu(A_1), \dots, Z_\mu(A_k)$ are (jointly) Gaussian random variables for all sets $A_j \in \mathcal{M}$, $\mu(A_j) < \infty$, $j = 1, \dots, k$.
(ii′) $EZ_\mu(A) = 0$, and $EZ_\mu(A)Z_\mu(B) = \mu(A \cap B)$ if $A, B \in \mathcal{M}$, $\mu(A) < \infty$, $\mu(B) < \infty$.

P. Major, *Multiple Wiener-Itô Integrals*, Lecture Notes in Mathematics 849, DOI 10.1007/978-3-319-02642-8_7,

It is not difficult to see that properties (i)–(iii) imply relations (i′) and (ii′). On the other hand, it is clear that (i′) and (ii′) imply (i) and (ii). To see that they also imply relation (iii) observe that under these conditions

$$E\left[Z_\mu\left(\bigcup_{j=1}^k A_j\right)-\sum_{j=1}^k Z_\mu(A_k)\right]^2=0$$

if $A_1,\dots,A_k$ are disjoint sets.

The second characterization of random orthogonal measures may help to show that for any measure space $(M,\mathcal{M},\mu)$ with a σ-finite measure μ there exists a Gaussian random orthogonal measure corresponding to the measure μ. The main point in checking this statement is the proof that for any sets $A_1,\dots,A_k\in\mathcal{M}$, $\mu(A_j)<\infty$, $1\le j\le k$, there exists a Gaussian random vector $(Z_\mu(A_1),\dots,Z_\mu(A_k))$, $EZ_\mu(A_j)=0$, with correlation $EZ_\mu(A_i)Z_\mu(A_j)=\mu(A_i\cap A_j)$ for all $1\le i,j\le k$. To prove this we have to show that the corresponding covariance matrix is really positive definite, i.e. $\sum\limits_{i,j} c_i\bar c_j\mu(A_i\cap A_j)\ge 0$ for an arbitrary vector $(c_1,\dots,c_k)$. But this follows from the observation $\sum\limits_{i,j} c_i\bar c_j\chi_{A_i\cap A_j}(x)=\sum\limits_{i,j} c_i\bar c_j\chi_{A_i}(x)\overline{\chi_{A_j}(x)}=\left|\sum\limits_i c_i\chi_{A_i}(x)\right|^2\ge 0$ for all $x\in M$, if we integrate this inequality with respect to the measure μ in the space M.

We define the real Hilbert spaces $\bar{\mathcal{K}}_\mu^n$, $n=1,2,\dots$. The space $\bar{\mathcal{K}}_\mu^n$ consists of the real-valued measurable functions over $(\underbrace{M\times\cdots\times M}_{n\text{ times}},\underbrace{\mathcal{M}\times\cdots\times\mathcal{M}}_{n\text{ times}})$ such that

$$\|f\|^2=\int|f(x_1,\dots,x_n)|^2\mu(dx_1)\dots\mu(dx_n)<\infty,$$

and the last formula defines the norm in $\bar{\mathcal{K}}_\mu^n$. Let $\mathcal{K}_\mu^n$ denote the subspace of $\bar{\mathcal{K}}_\mu^n$ consisting of the functions $f\in\bar{\mathcal{K}}_\mu^n$ such that

$$f(x_1,\dots,x_n)=f(x_{\pi(1)},\dots,x_{\pi(n)})\quad\text{for all }\pi\in\Pi_n.$$

Let the spaces $\bar{\mathcal{K}}_\mu^0$ and $\mathcal{K}_\mu^0$ consist of the real constants with the norm $\|c\|=|c|$. Finally we define the Fock space $\operatorname{Exp}\mathcal{K}_\mu$ which consists of the sequences $f=(f_0,f_1,\dots)$, $f_n\in\mathcal{K}_\mu^n$, $n=0,1,2,\dots$, such that

$$\|f\|^2=\sum_{n=0}^\infty\frac{1}{n!}\|f_n\|^2<\infty.$$

Given a random orthogonal measure Z_μ corresponding to μ, let us introduce the σ-algebra $\mathcal{F}=\sigma(Z_\mu(A)\colon A\in\mathcal{M},\ \mu(A)<\infty)$. Let $\mathcal{K}$ denote the real Hilbert space of square integrable random variables measurable with respect to the σ-algebra $\mathcal{F}$. Let $\mathcal{H}_{\le n}$ denote the subspace of $\mathcal{H}$ that is the closure of the linear space containing the polynomials of the random variables $Z_\mu(A)$ of order less than or equal to n. Let $\mathcal{H}_n$ be the orthogonal completion of $\mathcal{H}_{\le n-1}$ to $\mathcal{H}_{\le n}$. (The norm is defined as $\|\xi\|^2 = E\xi^2$ in these Hilbert spaces.)

The multiple Wiener–Itô integrals with respect to the random orthogonal measure Z_μ, to be defined below, give a unitary transformation from $\operatorname{Exp}\mathcal{H}_\mu$ to $\mathcal{H}$. We shall denote these integrals by $\int'$ to distinguish them from the Wiener–Itô integrals defined in Chap. 4.

First we define the class of simple functions $\hat{\bar{\mathcal{H}}}_\mu^n \subset \bar{\mathcal{H}}_\mu^n$. A function $f \in \bar{\mathcal{H}}_\mu^n$ is in $\hat{\bar{\mathcal{H}}}_\mu^n$ if there exists a finite system of disjoint sets $\Delta_1, \dots, \Delta_N$, with $\Delta_j \in \mathcal{M}$, $\mu(\Delta_j) < \infty$, $j = 1, \dots, N$, such that $f(x_1, \dots, x_n)$ is constant on the sets $\Delta_{j_1} \times \dots \times \Delta_{j_n}$ if the indices $j_1, \dots, j_n$ are disjoint, and $f(x_1, \dots, x_n)$ equals zero outside these sets. We define

$$\int' f(x_1, \dots, x_n) Z_\mu(dx_1) \dots Z_\mu(dx_n) = \sum f(x_{j_1}, \dots, x_{j_n}) Z_\mu(\Delta_{j_1}) \cdots Z_\mu(\Delta_{j_n})$$

for $f \in \hat{\bar{\mathcal{H}}}_\mu^n$, where $x_k \in \Delta_k$, $k = 1, \dots, N$. Here again, it can be seen with the help of the additivity property (iii) of the random orthogonal measure Z_μ that the above definition of the Wiener–Itô integral of simple functions is meaningful, although the simple function f does not determine uniquely the sets Δ_j appearing in this definition.

Let $\hat{\mathcal{H}}_\mu^n = \hat{\bar{\mathcal{H}}}_\mu^n \cap \mathcal{H}_\mu^n$. The random variables

$$I'_\mu(f) = \frac{1}{n!} \int' f(x_1, \dots, x_n) Z_\mu(dx_1) \dots Z_\mu(dx_n), \quad f \in \hat{\bar{\mathcal{H}}}_\mu^n,$$

have zero expectation, integrals of different order are orthogonal,

$$I'_\mu(f) = I'_\mu(\operatorname{Sym} f), \quad \text{and } \operatorname{Sym} f \in \hat{\mathcal{H}}_\mu^n \text{ if } f \in \hat{\bar{\mathcal{H}}}_\mu^n,$$

$$EI'_\mu(f)^2 \le \frac{1}{n!}\|f\|^2 \quad \text{if } f \in \hat{\bar{\mathcal{H}}}_\mu^n, \tag{7.1}$$

and (7.1) holds with equality if $f \in \hat{\mathcal{H}}_\mu^n$.

It can be seen that $\hat{\bar{\mathcal{H}}}_\mu^n$ is dense in $\bar{\mathcal{H}}_\mu^n$ in the $L_2(\mu^n)$ norm. (This is a statement analogous to Lemma 4.1, but its proof is simpler.) Hence relation (7.1) enables us to extend the definition of the n-fold Wiener–Itô integrals over $\bar{\mathcal{H}}_\mu^n$. All the above mentioned relations remain valid if $f \in \hat{\bar{\mathcal{H}}}_\mu^n$ is substituted by $f \in \bar{\mathcal{H}}_\mu^n$, and $f \in \hat{\mathcal{H}}_\mu^n$ is substituted by $f \in \mathcal{H}_\mu^n$. We formulate Itô's formula for these integrals. It can be proved similarly to Theorem 4.3 with the help of the diagram formula valid for the classical Wiener–Itô integrals studied in this chapter.

Theorem 7.1 (Itô's Formula). *Let $\varphi_1, \dots, \varphi_m$, $\varphi_j \in \mathcal{K}_\mu^1$ for all $1 \le j \le m$, be an orthonormal system in L_μ^2. Let some positive integers $j_1, \dots, j_m$ be given, put $j_1 + \dots + j_m = N$, and define for all $i = 1, \dots, N$*

$$g_i = \varphi_1 \text{ for } 1 \le i \le j_1, \quad \text{and } g_i = \varphi_s \quad \text{for } j_1 + \dots + j_{s-1} < i \le j_1 + \dots + j_s.$$

Then

$$\begin{aligned} H_{j_1}\left(\int' \varphi_1(x) Z_\mu(dx)\right) \cdots H_{j_m}\left(\int' \varphi_m(x) Z_\mu(dx)\right) \\ = \int' g_1(x_1) \cdots g_N(x_N)\, Z_\mu(dx_1) \dots Z_\mu(dx_N) \\ = \int' \operatorname{Sym}\left[g_1(x_1) \cdots g_N(x_N)\right] Z_\mu(dx_1) \dots Z_\mu(dx_N). \end{aligned}$$

(Let me remark that the diagram formula (Theorem 5.3) also remains valid for this integral if we replace $-x_j$ is by x_j and $G(dx_j)$ by $\mu(dx_j)$, $N - 2|\gamma| + 1 \le j \le N - |\gamma|$, in the definition of h_γ in formula (5.2).)

It can be seen with the help of Theorem 7.1 that the transformation

$$I'_\mu\colon \operatorname{Exp} \mathcal{K}_\mu \to \mathcal{K},$$

where $I'_\mu(f) = \sum\limits_{n=0}^{\infty} I'_\mu(f_n)$, $f = (f_0, f_1, \dots) \in \operatorname{Exp} \mathcal{K}_\mu$ is a unitary transformation, and so are the transformations $(n!)^{1/2} I'_\mu$ from $\mathcal{K}_\mu^n$ to $\mathcal{K}_n$.

Let us consider the special case $(M, \mathcal{M}, \mu) = (R^\nu, \mathcal{B}^\nu, \lambda)$, where λ denotes the Lebesgue measure in R^ν. A random orthogonal measure corresponding to λ is called the white noise. A random *spectral measure* corresponding to λ, when the Lebesgue measure is considered as the spectral measure of a generalized field, is also called a white noise. The next result, that can be considered as a random Plancherel formula, establishes a connection between the two types of Wiener–Itô integrals with respect to white noise.

Proposition 7.2. *Let $f = (f_0, f_1, \dots,) \in \operatorname{Exp} \mathcal{K}_\lambda$ be an element of the Fock space corresponding to the Lebesgue measure in the Euclidean space $(R^\nu, \mathcal{B}^\nu)$. Then $f' = (f'_0, f'_1, \dots,) \in \operatorname{Exp} \mathcal{H}_\lambda$ with the functions $f'_0 = f_0$ and $f'_n = (2\pi)^{-n\nu/2} \tilde{f}_n$, $n = 1, 2, \dots$, (where $\tilde{f}_n(u_1, \dots, u_n) = \int_{R^{n\nu}} e^{i(x,u)} f_n(x_1, \dots, x_n)\, dx_1 \dots dx_n$ with $x = (x_1, \dots, x_n)$ and $u = (u_1, \dots, u_n)$), and*

$$\begin{aligned} \sum_{n=0}^{\infty} \frac{1}{n!} \int' f_n(x_1, \dots, x_n)\, Z_\lambda(dx_1) \dots Z_\lambda(dx_n) \\ \stackrel{\Delta}{=} \sum_{n=0}^{\infty} \frac{1}{n!} \int f'_n(u_1, \dots, u_n)\, Z_\lambda(du_1) \dots Z_\lambda(du_n), \end{aligned}$$

where $Z_\lambda(dx)$ is a white noise as a random orthogonal measure, and $Z_\lambda(du)$ is a white noise as a random spectral measure.

Proof of Proposition 7.2. We have

$$(2\pi)^{-n\nu/2}\|\tilde{f}_n\|_{L^2_\lambda} = \|f_n\|_{L^2_\lambda},$$

hence $f' \in \operatorname{Exp}\mathscr{H}_\lambda$.

Let $\varphi_1, \varphi_2, \dots$ be a complete orthonormal system in L^2_λ. Then $\varphi'_1, \varphi'_2, \dots$ is also a complete orthonormal system in L^2_λ, and if

$$f_n(x_1, \dots, x_n) = \sum c_{j_1,\dots,j_n}\varphi_{j_1}(x_1)\cdots\varphi_{j_n}(x_n),$$

then

$$f'_n(u_1, \dots, u_n) = \sum c_{j_1,\dots,j_n}\varphi'_{j_1}(u_1)\cdots\varphi'_{j_n}(u_n).$$

Hence an application of Itô's formula for both types of integrals, (i.e. Theorems 4.3 and 7.1) imply Proposition 7.2. □

Finally we restrict ourselves to the case $\nu = 1$. We formulate a result which reflects a connection between multiple Wiener–Itô integrals and classical Itô integrals. Let $W(t)$, $a \le t \le b$, be a Wiener process, and let us define the random orthogonal measure $Z(dx)$ as

$$Z(A) = \int \chi_A(x)W(dx), \quad A \subset [a, b), \quad A \in \mathscr{B}^1.$$

Then we have the following

Proposition 7.3. *Let $f \in \mathscr{K}^n_{\lambda[a,b)}$, where $\lambda[a, b)$ denotes the Lebesgue measure on the interval $[a, b)$. Then*

$$\int' f(x_1, \dots, x_n)\, Z(dx_1)\dots Z(dx_n)$$
$$= n!\int_a^b \left(\int_a^{t_n}\left(\cdots\left(\int_a^{t_3}\left(\int_a^{t_2} f(t_1, \dots, t_n)W(dt_1)\right)W(dt_2)\right)\dots\right)W(dt_n)\right). \tag{7.2}$$

Proof of Proposition 7.3. Given a function $f \in \hat{\mathscr{K}}^n_{\lambda[a,b)}$, let the function $\hat{f}$ be defined as

$$\hat{f}(x_1, \dots, x_n) = \begin{cases} f(x_1, \dots, x_n) & \text{if } x_1 < x_2 < \dots < x_n \\ 0 & \text{otherwise.} \end{cases}$$

It is not difficult to check Proposition 7.3 for such a special function $f \in \hat{\mathcal{K}}^n_{\lambda[a,b)}$ for which the above defined function $\hat{f}$ is the indicator function of a rectangle of the form $\prod_{j=1}^{n}[a_j, b_j)$ with constants $a \le a_1 < b_1 < a_2 < b_2 < \cdots < a_n < b_n \le b$. Here we exploit the relation $I'(f) = n!I'(\hat{f})$. Beside this, we have to calculate the value of the right-hand side of formula (7.2) for these simple functions $f \in \hat{\mathcal{K}}^n_{\lambda[a,b)}$. A simple inductive argument shows that it equals $\prod_{j=1}^{n}[W(b_j) - W(a_j)]$ if $a \le a_1 < b_1 < a_2 < b_2 < \cdots < a_n < b_n \le b$, and it equals zero otherwise. Then a simple limiting procedure with the help of the approximation of general functions in $\mathcal{K}^n_{\lambda[a,b)}$ by the linear combinations of such functions proves Proposition 7.3 in the general case. □

As a consequence of Proposition 7.3 in the case $\nu = 1$ multiple Wiener–Itô integrals can be substituted by Itô integrals in the investigation of most problems. In the case $\nu = 2$ there is no simple definition of Itô integrals. On the other hand, no problem arises in generalizing the definition of multiple Wiener–Itô integrals to the case $\nu \ge 2$.

Chapter 8
Non-central Limit Theorems

In this chapter we investigate the problem formulated in Chap. 1, and we show how the technique of Wiener–Itô integrals can be applied for the investigation of such a problem. We restrict ourselves to the case of discrete random fields, although the case of generalized random fields can be discussed in almost the same way. We also present some generalizations of these results which can be proved in a similar way. But the proof of these more general results will be omitted. They can be found in [9]. First we recall the following

Definition 8A (Definition of Slowly Varying Functions). A function $L(t)$, $t \in [t_0, \infty)$, $t_0 > 0$, is said to be a slowly varying function (at infinity) if

$$\lim_{t\to\infty} \frac{L(st)}{L(t)} = 1 \quad \text{for all } s > 0.$$

We shall apply the following description of slowly varying functions.

Theorem 8A (Karamata's Theorem). *If a slowly varying function $L(t)$, $t \ge t_0$, with some $t_0 > 0$, is bounded on every finite interval, then it can be represented in the form*

$$L(t) = a(t) \exp\left\{ \int_{t_0}^{t} \frac{\varepsilon(s)}{s}\, ds \right\},$$

where $0 < \alpha < \nu$, $L(t)$, $t \ge 1$, is a slowly varying function, $t \to \infty$.

Let X_n, $n \in \mathbb{Z}_\nu$, be a stationary Gaussian field with expectation zero and a correlation function

$$r(n) = EX_0X_n = |n|^{-\alpha} a\left(\frac{n}{|n|}\right) L(|n|), \quad n \in \mathbb{Z}_\nu, \quad \text{if } n \ne (0,\dots,)), \tag{8.1}$$

P. Major, *Multiple Wiener-Itô Integrals*, Lecture Notes in Mathematics 849, DOI 10.1007/978-3-319-02642-8_8,

where $0 < \alpha < \nu$, $L(t)$, $t \ge 1$, is a slowly varying function, bounded in all finite intervals, and $a(t)$ is a continuous function on the unit sphere $\mathcal{S}_{\nu-1}$, satisfying the symmetry property $a(x) = a(-x)$ for all $x \in \mathcal{S}_{\nu-1}$. Let G denote the spectral measure of the field X_n, and let us define the measures G_N, $N = 1, 2, \ldots$, by the formula

$$G_N(A) = \frac{N^\alpha}{L(N)} G\left(\frac{A}{N}\right), \quad A \in \mathcal{B}^\nu, \quad N = 1, 2, \ldots. \tag{8.2}$$

Now we recall the definition of vague convergence of not necessarily finite measures on a Euclidean space.

Definition of Vague Convergence of Measures. *Let G_n, $n = 1, 2, \ldots$, be a sequence of locally finite measures over R^ν, i.e. let $G_n(A) < \infty$ for all measurable bounded sets A. We say that the sequence G_n vaguely converges to a locally finite measure G_0 on R^ν (in notation $G_n \xrightarrow{v} G_0$) if*

$$\lim_{n\to\infty} \int f(x)\, G_n(dx) = \int f(x)\, G_0(dx)$$

for all continuous functions f with a bounded support.

We formulate the following

Lemma 8.1. *Let G be the spectral measure of a stationary random field with a correlation function $r(n)$ of the form (8.1). Then the sequence of measures G_N defined in (8.2) tends vaguely to a locally finite measure G_0. The measure G_0 has the homogeneity property*

$$G_0(A) = t^{-\alpha} G_0(tA) \quad \text{for all } A \in \mathcal{B}^\nu \quad \text{and } t > 0, \tag{8.3}$$

and it satisfies the identity

$$2^\nu \int e^{i(t,x)} \prod_{j=1}^{\nu} \frac{1 - \cos x^{(j)}}{(x^{(j)})^2} G_0(dx)$$

$$= \int_{[-1,1]^\nu} (1 - |x^{(1)}|) \cdots (1 - |x^{(\nu)}|) \frac{a\left(\frac{x+t}{|x+t|}\right)}{|x+t|^\alpha}\, dx, \quad \text{for all } t \in R^\nu. \tag{8.4}$$

Remark. One may ask whether there are stationary random fields with correlation function satisfying relation (8.1), or more generally, how large the class of such random fields is. It can be proved that we get a correlation function of the form (8.1) with the help of a spectral measure with a spectral density of the form $g(u) = |u|^{\alpha-\nu} b(\frac{u}{|u|}) h(|u|)$, $u \in R^\nu$, where $b(\cdot)$ is a non-negative smooth function on the unit sphere $\{u\colon\ u \in R^\nu,\ |u| = 1\}$, and $h(u)$ is a non-negative smooth function on R^1 which does not disappear at the origin, and tends to zero at infinity sufficiently fast. The regularizing function $h(|u|)$ is needed in this formula to make the function $g(\cdot)$ integrable. Results of this type are studied in the theory of generalized functions.

At a heuristic level the class of spectral measures $G(\cdot)$ which determine a correlation function $r(\cdot)$ satisfying relation (8.1) can be described in the following way. They are such measures G for which the asymptotic identity $G(B_x) \sim Cx^\alpha$ holds with some constant $C > 0$ for the (small) balls $B_x = \{v\colon |v| \le x\}$ as $x \to 0$, and the effect of the singularities of the measure G outside the origin is in some sense small. At this heuristic level we disregarded the possibility of a factor $L(|x|)$ with a function $L(\cdot)$, slowly varying at the origin. Thus heuristically we can say that the asymptotically homogeneous behaviour $r(n) \sim Cn^{-\alpha}$ of the correlation function at infinity corresponds to the asymptotically homogeneous behaviour $G(B_x) \sim \bar{C}x^\alpha$ of the spectral measure G corresponding to it in the neighbourhood of zero together with some additional restrictions about the singularities of the spectral measure G outside zero which guarantee that their influence is not too strong. These considerations may help us to understand the content of one of the most important conditions in the subsequent Theorem 8.2.

We postpone the proof of Lemma 8.1 for a while.

Formulae (8.3) and (8.4) imply that the function $a(t)$ and the number α in the definition (8.1) of a correlation function $r(n)$ uniquely determine the measure G_0. Indeed, by formula (8.4) they determine the (finite) measure $\prod\limits_{j=1}^{\nu} \frac{1-\cos x^{(j)}}{(x^{(j)})^2} G_0(\,dx)$, since they determine its Fourier transform. Hence they also determine the measure G_0. (Formula (8.3) shows that G_0 is a locally finite measure.) Let us also remark that since $G_N(A) = G_N(-A)$ for all $N = 1, 2, \dots$ and $A \in \mathscr{B}^\nu$, the relation $G_0(A) = G_0(-A)$, $A \in \mathscr{B}^\nu$ also holds. These properties of the measure G_0 imply that it can be considered as the spectral measure of a generalized random field. Now we formulate

Theorem 8.2. *Let X_n, $n \in \mathbb{Z}_\nu$, be a stationary Gaussian field with a correlation function $r(n)$ satisfying relation (8.1). Let us define the stationary random field $\xi_j = H_k(X_j)$, $j \in \mathbb{Z}_\nu$, with some positive integer k, where $H_k(x)$ denotes the k-th Hermite polynomial with leading coefficient 1, and assume that the parameter α appearing in (8.1) satisfies the relation $0 < \alpha < \frac{\nu}{k}$ with this number k. If the random fields Z_n^N, $N = 1, 2, \dots$, $n \in \mathbb{Z}_\nu$, are defined by formula (1.1) with $A_N = N^{\nu - k\alpha/2} L(N)^{k/2}$ and the above defined $\xi_j = H_k(X_j)$, then their multi-dimensional distributions tend to those of the random field Z_n^*,*

$$Z_n^* = \int \tilde{\chi}_n(x_1 + \dots + x_k)\, Z_{G_0}(\,dx_1) \dots Z_{G_0}(\,dx_k), \qquad n \in \mathbb{Z}_\nu.$$

Here Z_{G_0} is a random spectral measure corresponding to the spectral measure G_0 which appeared in Lemma 8.1. The function $\tilde{\chi}_n(\cdot)$, $n = (n^{(1)}, \dots, n^{(\nu)})$, is (similarly to formula (6.2) Chap. 6) the Fourier transform of the indicator function of the ν-dimensional unit cube $\prod\limits_{p=1}^{\nu} [n^{(p)}, n^{(p)} + 1]$.

Remark. The condition that the correlation function $r(n)$ of the random field X_n, $n \in \mathbb{Z}_\nu$, satisfies formula (8.1) can be weakened. Theorem 8.2 and Lemma 8.1 remain valid if (8.1) is replaced by the slightly weaker condition

$$\lim_{T\to\infty} \sup_{n:\, n\in\mathbb{Z}_\nu,\, |n|\ge T} \frac{r(n)}{|n|^{-\alpha} a\left(\frac{n}{|n|}\right) L(|n|)} = 1,$$

where $0 < \alpha < \nu$, $L(t)$, $t \ge 1$, is a slowly varying function, bounded in all finite intervals, and $a(t)$ is a continuous function on the unit sphere $\mathcal{S}_{\nu-1}$, satisfying the symmetry property $a(x) = a(-x)$ for all $x \in \mathcal{S}_{\nu-1}$.

First we explain why the choice of the normalizing constant A_N in Theorem 8.2 was natural, then we explain the ideas of the proof, finally we work out the details.

It can be shown, for instance with the help of Corollary 5.5, that $EH_k(\xi)H_k(\eta) = E{:}\xi^k{:}\,{:}\eta^k{:} = k!(E\xi\eta)^k$ for a Gaussian random vector (ξ, η) with $E\xi = E\eta = 0$ and $E\xi^2 = E\eta^2 = 1$. Hence

$$E(Z_n^N)^2 = \frac{k!}{A_N^2} \sum_{j,l\in B_0^N} r(j-l)^k \sim \frac{k!}{A_N^2} \sum_{j,l\in B_0^N} |j-l|^{-k\alpha} a^k\left(\frac{j-l}{|j-l|}\right) L(|j-l|)^k,$$

with the set B_0^N introduced after formula (1.1). Some calculation with the help of the above formula shows that with our choice of A_N the expectation $E(Z_n^N)^2$ is separated both from zero and infinity, therefore this is the natural norming factor. In this calculation we have to exploit the condition $k\alpha < \nu$, which implies that in the sum expressing $E(Z_n^N)^2$ those terms are dominant for which $j - l$ is relatively large, more explicitly which are of order N. There are const. $N^{2\nu}$ such terms.

The field ξ_n, $n \in \mathbb{Z}_\nu$, is subordinated to the Gaussian field X_n. It is natural to write up its canonical representation defined in Chap. 6, and to express Z_n^N via multiple Wiener–Itô integrals. Itô's formula yields the relation

$$\xi_j = H_k\left(\int e^{i(j,x)} Z_G(dx)\right) = \int e^{i(j,x_1+\cdots+x_k)} Z_G(dx_1)\dots Z_G(dx_k),$$

where Z_G is the random spectral measure adapted to the random field X_n. Then

$$\begin{aligned} Z_n^N &= \frac{1}{A_N} \sum_{j\in B_n^N} \int e^{i(j,x_1+\cdots+x_k)} Z_G(dx_1)\dots Z_G(dx_k) \\ &= \frac{1}{A_N} \int e^{i(Nn,x_1+\cdots+x_k)} \prod_{j=1}^{\nu} \frac{e^{iN(x_1^{(j)}+\cdots+x_k^{(j)})}-1}{e^{i(x_1^{(j)}+\cdots+x_k^{(j)})}-1} Z_G(dx_1)\dots Z_G(dx_k). \end{aligned}$$

Let us make the substitution $y_j = Nx_j$, $j = 1, \dots, k$, in the last formula, and let us rewrite it in a form resembling formula (6.8). To this end, let us introduce the measures G_N defined in (8.2). By Lemma 4.6 we can write

$$Z_n^N \stackrel{\Delta}{=} \int f_N(y_1,\dots,y_k)\tilde{\chi}_n(y_1+\cdots+y_k)\, Z_{G_N}(dy_1)\dots Z_{G_N}(dy_k)$$

with

$$f_N(y_1,\dots,y_k)=\prod_{j=1}^{\nu}\frac{i(y_1^{(j)}+\cdots+y_k^{(j)})}{\left(\exp\left\{i\frac{1}{N}(y_1^{(j)}+\cdots+y_k^{(j)})\right\}-1\right)N}, \tag{8.5}$$

where $\tilde{\chi}_n(\cdot)$ is the Fourier transform of the indicator function of the unit cube $\prod_{j=1}^{\nu}[n^{(j)}, n^{(j)}+1)$. (It follows from Lemma 8B formulated below and the Fubini theorem that the set, where the denominator of the function f_N disappears, i.e. the set where $y_1^{(j)}+\cdots+y_k^{(j)}=2lN\pi$ with some integer $l\neq 0$ and $1\leq j\leq\nu$ has zero $G_N\times\cdots\times G_N$ measure. This means that the functions f_N are well defined.) The functions f_N tend to 1 uniformly in all bounded regions, and the measures G_N tend vaguely to G_0 as $N\to\infty$ by Lemma 8.1. These relations suggest the following limiting procedure. The limit of Z_n^N can be obtained by substituting f_N with 1 and G_N with G_0 in the Wiener–Itô integral expressing Z_n^N. We want to justify this formal limiting procedure. For this we have to show that the Wiener–Itô integral expressing Z_n^N is essentially concentrated in a large bounded region independent of N. The L_2-isomorphism of Wiener–Itô integrals can help us in showing that. We shall formulate a result in Lemma 8.3 which is a useful tool for the justification of the above limiting procedure.

Before formulating this lemma we make a small digression. It was explained that Wiener–Itô integrals can be defined also with respect to random stationary fields Z_G adapted to a stationary Gaussian random field whose spectral measure G may have atoms, and we can work with them similarly as in the case of non-atomic spectral measures. Here a lemma will be proved which shows that in the proof of Theorem 8.2 we do not need this observation, because if the correlation function of the random field satisfies (8.1), then its spectral measure is non-atomic. Moreover, the measure G has an additional property which guarantees that the function $f_N(y_1,\dots,y_n)$ introduced in (8.5) can be defined in the space $R^{k\nu}$ with the product measure $G_N\times\cdots\times G_N$.

Lemma 8B. *Let the correlation function of a stationary random field X_n, $n\in\mathbb{Z}_\nu$, satisfy the relation $r(n)\leq A|n|^{-\alpha}$ with some $A>0$ and $\alpha>0$ for all $n\in\mathbb{Z}_\nu$, $n\neq 0$. Then its spectral measure G is non-atomic. Moreover, the hyperplanes $x^{(j)}=t$ have zero G measure for all $1\leq j\leq\nu$ and $t\in R^1$.*

Proof of Lemma 8B. Lemma 8B clearly holds if $\alpha>\nu$, because in this case the spectral measure G has even a density function $g(x)=\sum_{n\in\mathbb{Z}_\nu}e^{-i(n,x)}r(n)$. On the other hand, the p-fold convolution of the spectral measure G with itself (on the torus $R^\nu/2\pi\mathbb{Z}_\nu$) has Fourier transform, $r(n)^p$, $n\in Z^\nu$, and as a consequence in the case $p>\frac{\nu}{\alpha}$ this measure is non-atomic. Hence it is enough to show that if the convolution

$G * G$ is a non-atomic measure, then the measure G is also non-atomic. But this is obvious, because if there were a point $x \in R^\nu/2\pi\mathbb{Z}_\nu$ such that $G(\{x\}) > 0$, then the relation $G * G(\{x + x\}) > 0$ would hold, and this is a contradiction. (Here addition is taken on the torus.) The second statement of the lemma can be proved with some small modifications of the previous proof, by reducing it to the one-dimensional case. □

Now we formulate a result that helps us in carrying out some limiting procedures.

Lemma 8.3. *Let G_N, $N = 1, 2, \dots$, be a sequence of non-atomic spectral measures on R^ν tending vaguely to a non-atomic spectral measure G_0. Let a sequence of measurable functions $K_N = K_N(x_1, \dots, x_k)$, $N = 0, 1, 2, \dots$, be given such that $K_N \in \bar{\mathcal{H}}^k_{G_N}$ for $N = 1, 2, \dots$. Assume further that these functions satisfy the following properties: For all $\varepsilon > 0$ there exist some constants $A = A(\varepsilon) > 0$ and $N_0 = N_0(\varepsilon) > 0$ and finitely many rectangles $P_1, \dots, P_M$ with some cardinality $M = M(\varepsilon)$ on $R^{k\nu}$ such that the following conditions (a) and (b) formulated below with the help of these numbers and rectangles are satisfied. (We call a set $P \in \mathcal{B}^{k\nu}$ a rectangle if it can be written in the form $P = L_1 \times \dots \times L_k$ with some bounded open sets $L_s \in \mathcal{B}^\nu$, $1 \le s \le k$, with boundaries ∂L_s of zero G_0 measure, i.e. $G_0(\partial L_s) = 0$ for all $1 \le s \le k$.)*

(a) The function K_0 is continuous on the set $B = [-A, A]^{k\nu} \setminus \bigcup_{j=1}^{M} P_j$, and $K_N \to K_0$ uniformly on the set B as $N \to \infty$. Besides, the hyperplanes $x_p = \pm A$ have zero G_0 measure for all $1 \le p \le \nu$.

(b) $\int_{R^{k\nu}\setminus B} |K_N(x_1, \dots, x_k)|^2 G_N(dx_1) \dots G_N(dx_k) < \frac{\varepsilon^3}{k!}$ if $N = 0$ or $N \ge N_0$, and $K_0(-x_1, \dots, -x_k) = \overline{K_0(x_1, \dots, x_k)}$ for all $(x_1, \dots, x_k) \in R^{k\nu}$.

Then $K_0 \in \bar{\mathcal{H}}^k_{G_0}$, and

$$\int K_N(x_1, \dots, x_k) Z_{G_N}(dx_1) \dots Z_{G_N}(dx_k) \xrightarrow{\mathcal{D}} \int K_0(x_1, \dots, x_k) Z_{G_0}(dx_1) \dots Z_{G_0}(dx_k)$$

as $N \to \infty$, where $\xrightarrow{\mathcal{D}}$ denotes convergence in distribution.

Remark. In the proof of Theorem 8.2 or of its generalization Theorem 8.2′ formulated later a simpler version of Lemma 8.3 with a simpler proof would suffice. We could work with such a version where the rectangles P_j do not appear. We formulated this somewhat more complicated result, because it can be applied in the proof of more general theorems, where the limit is given by such a Wiener–Itô integral whose kernel function may have discontinuities. Thus it seemed to be better to present such a result even if its proof is more complicated. The proof applies some arguments of Lemma 4.1. To work out the details it turned out to be useful to introduce some metric in the space of probability measures which metricizes weak convergence. Although this may look a bit too technical, it made possible to carry out some arguments in a natural way. We can tell with the help of this notion when two probability measures are close to each other.

Proof of Lemma 8.3. Conditions (a) and (b) obviously imply that

$$\int |K_0(x_1,\dots,x_k)|^2\, G_0(\,dx_1)\dots G_0(\,dx_k) < \infty,$$

hence $K_0 \in \bar{\mathcal{H}}^k_{G_0}$. Let us fix an $\varepsilon > 0$, and let us choose some $A > 0$, $N_0 > 0$ and rectangles $P_1,\dots,P_M$ which satisfy conditions (a) and (b) with this ε. Then

$$\begin{aligned} &E\left[\int [1-\chi_B(x_1,\dots,x_k)]K_N(x_1,\dots,x_k)\, Z_{G_N}(\,dx_1)\dots Z_{G_N}(\,dx_k)\right]^2 \\ &\quad\le k!\int_{R^{k\nu}\setminus B} |K_N(x_1,\dots,x_k)|^2 G_N(\,dx_1)\dots G_N(\,dx_k) < \varepsilon^3 \end{aligned} \tag{8.6}$$

for $N = 0$ or $N > N_0$, where χ_B denotes the indicator function of the set B introduced in the formulation of condition (a).

Since $B \subset [-A, A]^{k\nu}$, and $G_N \xrightarrow{v} G_0$, hence $G_N \times \dots \times G_N(B) < C(A)$ with an appropriate constant $C(A) < \infty$ for all $N = 0, 1, \dots$. Because of this estimate and the uniform convergence $K_N \to K_0$ on the set B we have

$$\begin{aligned} &E\left[\int (K_N(x_1,\dots,x_k) - K_0(x_1,\dots,x_k))\chi_B(x_1,\dots,x_k)\, Z_{G_N}(\,dx_1)\dots Z_{G_N}(\,dx_k)\right]^2 \\ &\quad\le k!\int_B |K_N(x_1,\dots,x_k) - K_0(x_1,\dots,x_k)|^2\, G_N(\,dx_1)\dots G_N(\,dx_k) < \varepsilon^3 \end{aligned} \tag{8.7}$$

for $N > N_1$ with some $N_1 = N_1(A,\varepsilon)$.

First we reduce the proof of Lemma 8.3 to the proof of the relation

$$\begin{aligned} &\int K_0(x_1,\dots,x_k)\chi_B(x_1,\dots,x_k)\, Z_{G_N}(\,dx_1)\dots Z_{G_N}(\,dx_k) \\ &\quad\xrightarrow{\mathcal{D}} \int K_0(x_1,\dots,x_k)\chi_B(x_1,\dots,x_k)\, Z_{G_0}(\,dx_1)\dots Z_{G_0}(\,dx_k) \end{aligned} \tag{8.8}$$

with the help of formulas (8.6) and (8.7), and then we shall prove (8.8). It is simpler to carry out this reduction with the help of some metric on the space of probability measure which induces weak convergence in this space. Hence I recall some classical notions and results about convergence of probability measures on a metric space which will be useful in our considerations.

Definition of Prokhorov Metric, and Its Properties. *Given a separable metric space (X,ρ) with some metric ρ let $\mathcal{S}$ denote the space of probability measures on it. The Prokhorov metric ρ_P is the metric in the space $\mathcal{S}$ defined by the formula $\rho_P(\mu,\nu) = \inf\{\varepsilon\colon\ \mu(A) \le \nu(A^\varepsilon) + \varepsilon$ for all $A \in \mathcal{A}\}$ for two probability measures*

$\mu, \nu \in \mathcal{S}$, where $A^\varepsilon = \{x\colon \rho(x, A) < \varepsilon\}$. The above defined ρ_P is really a metric on $\mathcal{S}$ (in particular, $\rho_P(\mu, \nu) = \rho_P(\nu, \mu)$) which metricizes the weak convergence of probability measures in the metric space (X, ρ), i.e. $\mu_N \stackrel{w}{\to} \mu_0$ for a sequence of probability measures $N = 0, 1, 2, \dots$ if and only if $\lim_{N\to\infty} \rho_P(\mu_N, \mu_0) = 0$.

The results formulated in this definition can be found e.g. in [13]. Let us also recall the definition of weak converges of probability measures on a metric space.

Definition of Weak Convergence of Probability Measures on a Metric Space. *A sequence of probability measures μ_n, $n = 1, 2, \dots$, on a metric space (X, ρ) converges weakly to a probability measure μ on this space, (in notation $\mu_n \stackrel{w}{\to} \mu$) if $\lim_{n\to\infty} \int f(x)\mu_n(dx) \to \int f(x)\mu(dx)$ for all continuous and bounded functions on the space (X, ρ).*

I formulated the above result for probability measures in a general metric space, but I shall work on the real line. Given a random variable ξ let $\mu(\xi)$ denote its distribution. Let us remark that the convergence $\xi_N \stackrel{\mathcal{D}}{\to} \xi_0$ as $N \to \infty$ of a sequence of random variables, $\xi_0, \xi_1, \xi_2, \dots$ is equivalent to the statement $\mu(\xi_N) \stackrel{w}{\to} \mu(\xi_0)$ or $\rho_P(\mu(\xi_N), \mu(\xi_0)) \to 0$ as $N \to \infty$. Hence by putting $\xi_N = k! I_{G_N}(K_N(x_1, \dots, x_k))$, $N = 0, 1, 2, \dots$ we can reformulate the statement of Lemma 8.3 in the following way. For all $\varepsilon > 0$ there exists some index $N_0' = N_0'(\varepsilon)$ such that $\rho_P(\mu(\xi_N), \mu(\xi_0)) \le 4\varepsilon$ for all $N \ge N_0'$.

To reduce the proof of Lemma 8.3 to that of formula (8.8) first we show that for three random variables ξ, $\bar\xi$ and η such that $P(|\eta| \ge \varepsilon) \le \varepsilon$ the inequality

$$\rho_P(\mu(\xi + \eta), \mu(\bar\xi)) \le \rho_P(\mu(\xi), \mu(\bar\xi)) + \varepsilon \tag{8.9}$$

holds.

As ρ_P is a metric we can write $\rho_P(\mu(\xi + \eta), \mu(\bar\xi)) \le \rho_P(\mu(\xi + \eta), \mu(\xi)) + \rho_P(\mu(\xi), \mu(\bar\xi))$, and to prove (8.9) it is enough to show that $\rho_P(\mu(\xi+\eta), \mu(\xi)) \le \varepsilon$ if $P(|\eta| \ge \varepsilon) \le \varepsilon$.

This inequality holds, since $\{\omega\colon \xi(\omega) \in A\} \subset \{\omega\colon \xi(\omega) + \eta(\omega) \in A^\varepsilon\} \cup \{\omega\colon |\eta(\omega)| \ge \varepsilon\}$, and as a consequence $P(\xi \in A) \le P(\xi+\eta \in A^\varepsilon) + P(|\eta| \ge \varepsilon) \le P(\xi + \eta \in A^\varepsilon) + \varepsilon$ for any set $A \in \mathcal{B}_1$ if $P(|\eta| \ge \varepsilon) \le \varepsilon$. By the definition of the Prokhorov metric this means that the desired inequality holds.

Put

$$\begin{aligned}
\xi_N^{(1)} &= k! I_{G_N}(K_0(x_1, \dots, x_k)\chi_B(x_1, \dots, x_k)),\\
\xi_N^{(2)} &= k! I_{G_N}(K_N(x_1, \dots, x_k) - K_0(x_1, \dots, x_k))\chi_B(x_1, \dots, x_k)),\\
\xi_N^{(3)} &= k! I_{G_N}(1 - \chi_B(x_1, \dots, x_k))K_N(x_1, \dots, x_k)),\\
\xi_N &= k! I_{G_N}(K_N)
\end{aligned}$$

for all $N = 0, 1, 2, \dots$. With this notation it follows from relation (8.8) and the fact that the Prokhorov metric metricizes the weak convergence that

$$\rho_P(\mu(\xi_N^{(1)}), \mu(\xi_0^{(1)})) \le \varepsilon \quad \text{if } N \ge N_1'(\varepsilon)$$

with some threshold index $N_1'(\varepsilon)$. Formulas (8.6) and (8.7) together with the Chebishev inequality imply that $P(|\xi_N^{(2)}| \ge \varepsilon) \le \varepsilon$ and $P(|\xi_N^{(3)}| \ge \varepsilon) \le \varepsilon$ if $N \ge N_2'(\varepsilon)$ or $N = 0$ with some threshold index $N_2'(\varepsilon)$. Besides, we have $\xi_0 = \xi_0^{(1)} + \xi_0^{(3)}$ and $\xi_N = \xi_N^{(1)} + \xi_N^{(2)} + \xi_N^{(3)}$ for $N = 1, 2, \dots$. The above mentioned properties of the random variables we considered together with relation (8.9) imply that

$$\begin{aligned}
\rho_P(\mu(\xi_N), \mu(\xi_0)) &= \rho_P(\mu(\xi_N^{(1)} + \xi_N^{(2)} + \xi_N^{(3)}), \mu(\xi_0^{(1)} + \xi_0^{(3)})) \\
&\le \rho_P(\mu(\xi_N^{(1)} + \xi_N^{(2)} + \xi_N^{(3)}), \mu(\xi_0^{(1)})) + \varepsilon \\
&\le \rho_P(\mu(\xi_N^{(1)} + \xi_N^{(2)}), \mu(\xi_0^{(1)})) + 2\varepsilon \\
&\le \rho_P(\mu(\xi_N^{(1)}), \mu(\xi_0^{(1)})) + 3\varepsilon \le 4\varepsilon
\end{aligned}$$

if $N \ge N_0'(\varepsilon) = \max(N_1'(\varepsilon), N_2'(\varepsilon))$. Hence Lemma 8.3 follows from (8.8).

To prove (8.8) we will show that $K_0(x_1, \dots, x_k)\chi_B(x_1, \dots, x_k)$ can be well approximated by simple functions from $\hat{\bar{\mathscr{H}}}_{G_0}^k$ in the following way. For all $\varepsilon' > 0$ there exists a simple function $f_{\varepsilon'} \in \hat{\bar{\mathscr{H}}}_{G_0}^k$ such that

$$E \int (K_0(x_1, \dots, x_k)\chi_B(x_1, \dots, x_k) - f_{\varepsilon'}(x_1, \dots, x_k))^2 G_0(dx_1) \dots G_0(dx_k) \le \frac{\varepsilon'^3}{k!} \tag{8.10}$$

and also

$$E \int (K_0(x_1, \dots, x_k)\chi_B(x_1, \dots, x_k) - f_{\varepsilon'}(x_1, \dots, x_k))^2 G_N(dx_1) \dots G_N(dx_k) \le \frac{\varepsilon'^3}{k!} \tag{8.11}$$

if $N \ge N_0$ with some threshold index $N_0 = N_0(\varepsilon', K_0(\cdot)\chi_B(\cdot))$. Moreover, this simple function $f_{\varepsilon'}$ can be chosen in such a way that it is adapted to such a regular system $\mathscr{D} = \{\Delta_j,\ j = \pm 1, \dots, \pm M\}$ whose elements have boundaries with zero G_0 measure, i.e. $G_0(\partial \Delta_j) = 0$ for all $1 \le |j| \le M$.

To prove (8.8) with the help of these estimates first we show that this function $f_{\varepsilon'} \in \hat{\bar{\mathscr{H}}}_{G_0}^k$ satisfies the relation

$$\int f_{\varepsilon'}(x_1, \dots, x_k) Z_{G_N}(dx_1) \dots Z_{G_N}(dx_k) \xrightarrow{\mathscr{D}} \int f_{\varepsilon'}(x_1, \dots, x_k) Z_{G_0}(dx_1) \dots Z_{G_0}(dx_k) \tag{8.12}$$

as $N \to \infty$. In the proof of (8.12) we exploit that we can take such a regular system $\mathscr{D} = \{\Delta_j,\ j = \pm 1, \dots, \pm M\}$ to which the function $f_{\varepsilon'} \in \hat{\bar{\mathscr{H}}}_{G_0}^k$ is adapted and which has the property $G_0(\partial\Delta_j) = 0$ for all $j = \pm 1, \dots, \pm M$. Besides, the spectral measures G_N are such that $G_N \xrightarrow{v} G_0$. Hence the (Gaussian) random vectors $(Z_{G_N}(\Delta_j),\ j = \pm 1, \dots, \pm M)$ converge in distribution to the (Gaussian) random vector $(Z_{G_0}(\Delta_j),\ j = \pm 1, \dots, \pm M)$ as $N \to \infty$. The same can be told about such random variables that we get by putting the arguments of these random vectors to a continuous function (of $2M$ variables). Since the integrals in (8.12) are polynomials of these random vectors, we can apply these results for them, and they imply relation (8.12).

Put

$$K_0(x_1, \dots, x_k)\chi_B(x_1, \dots, x_k) - f_{\varepsilon'}(x_1, \dots, x_k) = h_0(x_1, \dots, x_k). \tag{8.13}$$

By relations (8.10), (8.11) and the Chebishev inequality $P(|k!I_{G_0}(h_0)| \ge \varepsilon') \le \varepsilon'$ and $P(|k!I_{G_N}(h_0) \ge \varepsilon') \le \varepsilon'$ if $N \ge N_0$. Since $I_{G_N}(K_N(x_1, \dots, x_k)\chi_B(x_1, \dots, x_k)) = I_{G_N}(f_{\varepsilon'}(x_1, \dots, x_k) + (h_0(x_1, \dots, x_k)),\ N = 0, 1, 2, \dots,$ the above relations together with formulas (8.12) and (8.9) (with the number ε' instead of ε) imply that

$$\begin{aligned}
&\limsup_{N\to\infty} \rho_P(\mu(k!I_{G_N}(K_0(\cdot)\chi_B(\cdot))), \mu(k!I_{G_0}(K_0(\cdot)\chi_B(\cdot))))\\
&\quad= \limsup_{N\to\infty} \rho_P(\mu(k!I_{G_N}(f_{\varepsilon'}(\cdot) + h_0(\cdot))), \mu(k!I_{G_0}(f_{\varepsilon'}(\cdot) + h_0(\cdot))))\\
&\quad= \limsup_{N\to\infty} \rho_P(\mu(k!I_{G_N}(f_{\varepsilon'}(\cdot)) + k!I_{G_N}(h_0(\cdot))), \mu(k!I_{G_0}(f_{\varepsilon'}(\cdot)) + k!I_{G_0}(h_0(\cdot))))\\
&\quad\le \limsup_{N\to\infty} \rho_P(\mu(k!I_{G_N}(f_{\varepsilon'}(\cdot))), \mu(k!I_{G_0}(f_{\varepsilon'}(\cdot)))) + 2\varepsilon' = 2\varepsilon'.
\end{aligned}$$

Since this inequality holds for all $\varepsilon' > 0$ this implies relation (8.8). To complete the proof of Lemma 8.3 we have to justify relations (8.10) and (8.11).

Relation (8.10) is actually a version of Lemma 4.1, but it states a slightly stronger approximation result under the conditions of Lemma 8.3. The statement that for all ε' the function $K_0(\cdot)\chi_B(\cdot)$ can be approximated with a simple function $f_{\varepsilon'}(x_1, \dots, x_k)$ which satisfies (8.10) agrees with Lemma 4.1. But now we want to choose such a simple function $f_{\varepsilon'}$ which is adapted to a regular system $\mathcal{D} = \{\Delta_j,\ j = \pm 1, \dots, \pm M\}$ with such elements that have the additional property $G_0(\partial\Delta_j) = 0$ for all indices j. A function $f_{\varepsilon'}$ with these properties can be constructed by means of a slight modification of the proof of Lemma 4.1. We exploit that in the present case the function $K_0(\cdot)\chi_B(\cdot)$ is almost everywhere continuous with respect to the product measure $G_0^k = \underbrace{G_0 \times \cdots \times G_0}_{k \text{ times}}$. This property is needed in the first step of the construction, where we reduce the approximation result we want to prove to a slightly modified version of *Statement A*.

In this modified version of *Statement A* we want to find a good approximation of the indicator function of such sets A which satisfies not only the properties demanded in *Statement A*, but also the identities $G_0(\partial A) = 0$ and $G_0(\partial A_1) = 0$ hold for them. On the other hand, we demand the identity $G_0(\partial B) = 0$ also for the set B whose indicator function is the approximating function in *Statement A*. To carry out the reduction, needed in this case we approximate the function $K_0(\cdot)\chi_B(\cdot)$ with such an elementary function (a function taking finitely many values) whose level sets have boundaries with zero $G_0^k = G_0 \times \cdots \times G_0$ measure. This is possible, since the boundaries of these level sets consist of such points where either the function $K_0(\cdot)\chi_B(\cdot)$ takes the value from an appropriately chosen finite set, or this function is discontinuous. At this point we exploit that the function $K_0(\cdot)\chi_B(\cdot)$ is almost everywhere continuous with respect to the measure G_0.

To complete the reduction of the proof of (8.10) to the new version of *Statement A* we still have to show that if the set A can be written in the form $A = A_1 \cup (-A_1)$ such that $A_1 \cap (-A_1) = \emptyset$, and $G_0^k(\partial A_1) = 0$, then for all $\eta > 0$ there is some $\bar{A}_1 = \bar{A}_1(\eta) \subset A_1$ such that $G_0^k(A \setminus (\bar{A}_1 \cup (-\bar{A}_1)) \le \eta$, $\rho(\bar{A}_1, -\bar{A}_1) > 0$, and $G_0^k(\partial \bar{A}_1) = 0$. Indeed, there is a compact set $K \subset A_1$ such that $G^k(A_1 \subset K) \le \frac{\eta}{2}$. Then also the relation $\rho(K, -K) = \delta > 0$ holds. By the Heine–Borel theorem we can find an open set G such that $K \subset G \subset K^{\delta/3}$ with $K^{\delta/3} = \{x\colon \rho(x, K) < \frac{\delta}{3}\}$, and $G_0^k(\partial G) = 0$. Then the set $\bar{A}_1 = A_1 \cap G$ satisfies the desired properties.

After making the reduction of the result we want to prove to this modified version of *Statement A* we can follow the construction of Lemma 4.1, but we choose in each step sets with zero $G_0 \times \cdots \times G_0$ boundary.

A more careful analysis shows that the function constructed in such a way satisfies also (8.11) for $N \ge N_0$ with a sufficiently large threshold index N_0. Here we exploit that $G_N \xrightarrow{v} G_0$. This enables us to show that the estimates we need in the construction hold not only with respect to the spectral measure G_0 but also with respect to the spectral measures G_N with a sufficiently large index N. We can get another explanation of the estimate (8.11) by exploiting that the function $h_0(x_1, \dots, x_k)$ defined in (8.13) is almost everywhere continuous with respect to the measure $G_0 \times \cdots \times G_0$. It can be shown that the vague convergence has similar properties as the weak convergence. In particular, the above mentioned almost everywhere continuity implies that

$$\lim_{N\to\infty} \int h_0(x_1, \dots, x_k) G_N(dx_1) \dots G_N(dx_k) = \int h_0(x_1, \dots, x_k) G_0(dx_1) \dots G_0(dx_k).$$

□

Remark. In Lemma 8.3 we proved the convergence of Wiener–Itô integrals with respect to random spectral measures Z_{G_N} corresponding to spectral measures G_N on the Euclidean space R^ν under appropriate conditions. There is a natural version of this result which we get by considering Wiener–Itô integrals $k! I_{G_N}(K_N)$ on the torus of size $2C_N\pi$ with some numbers $C_N \to \infty$ as $N \to \infty$. To find a good formulation of the result in this case observe that the torus $R^\nu / 2\pi\mathbb{Z}_\nu$ can be identified with the set $[-C_N\pi, C_N\pi)^\nu \subset R^\nu$ in a natural way. This identification enables us to consider the spectral measure G_N as a measure on $[-C_N\pi, C_N\pi)^\nu$ and the function K_N as a function on this set, which can be extended to a function on R^ν, periodic in all of its coordinates with periodicity $2\pi C_N$. With such a notation we demand in this version of Lemma 8.3 that $G_N \xrightarrow{v} G_0$, and conditions (a) and (b) hold with these (non-atomic) measures G_N and functions K_N. This version of Lemma 8.3 can be proved in almost the same way. We can reduce its proof to the verification of formula (8.8), and after this it has no importance whether we work in R^ν or in $[-C_N\pi, C_N\pi)^\nu$.

Now we turn to the proof of Theorem 8.2.

Proof of Theorem 8.2. We want to prove that for all positive integers p, real numbers $c_1, \dots, c_p$ and $n_l \in \mathbb{Z}_\nu$, $l = 1, \dots, p$,

$$\sum_{l=1}^{p} c_l Z_{n_l}^N \xrightarrow{\mathcal{D}} \sum_{l=1}^{p} c_l Z_{n_l}^*,$$

since this relation also implies the convergence of the multi-dimensional distributions. Applying the same calculation as before we get with the help of Lemma 4.6 that

$$\sum_{l=1}^{p} c_l Z_{n_l}^N = \frac{1}{A_N} \sum_{l=1}^{p} c_l \int \sum_{j \in B_{n_l}^N} e^{i(j,x_1+\cdots+x_k)} Z_G(dx_1) \dots Z_G(dx_k),$$

and

$$\sum_{l=1}^{p} c_l Z_{n_l}^N \stackrel{\Delta}{=} \int K_N(x_1, \dots, x_k)\, Z_{G_N}(dx_1) \dots Z_{G_N}(dx_k)$$

with

$$K_N(x_1, \dots, x_k) = \frac{1}{N^\nu} \sum_{l=1}^{p} c_l \sum_{j \in B_{n_l}^N} \exp\left\{ i \left(\frac{j}{N}, x_1 + \cdots + x_k \right) \right\}$$

$$= f_N(x_1, \dots, x_k) \sum_{l=1}^{p} c_l \tilde{\chi}_{n_l}(x_1 + \cdots + x_k). \tag{8.14}$$

with the function f_N defined in (8.5) and the measure G_N defined in (8.2), The function $\tilde{\chi}_n(\cdot)$ denotes again the Fourier transform of the indicator function of the unit cube $\prod_{j=1}^{\nu} [n^{(j)}, n^{(j)} + 1)$, $n = (n^{(1)}, \dots n^{(\nu)})$.

Let us define the function

$$K_0(x_1, \dots, x_k) = \sum_{l=1}^{p} c_l \tilde{\chi}_{n_l}(x_1 + \cdots + x_k)$$

and the measures μ_N on $R^{k\nu}$ by the formula

$$\mu_N(A) = \int_A |K_N(x_1, \dots, x_k)|^2 G_N(dx_1) \dots G_N(dx_k),$$
$$A \in \mathscr{B}^{k\nu} \text{ and } N = 0, 1, \dots. \tag{8.15}$$

In the case $N = 0$ G_0 is the vague limit of the measures G_N.

We prove Theorem 8.2 by showing that Lemma 8.3 can be applied with these spectral measures G_N and functions K_N. (We choose no exceptional rectangles P_j in this application of Lemma 8.3.) Since $G_N \xrightarrow{v} G_0$, and $K_N \to K_0$ uniformly in all bounded regions in $R^{k\nu}$, it is enough to show, beside the proof of Lemma 8.1, that the measures μ_N, $N = 1, 2, \dots$, tend weakly to the (necessarily finite) measure μ_0 which is also defined in (8.15), (in notation $\mu_N \xrightarrow{w} \mu_0$), i.e. $\int f(x)\mu_N(dx) \to \int f(x)\mu_0(dx)$ for all continuous and bounded functions f on $R^{k\nu}$. Then this convergence implies condition (b) in Lemma 8.3. Moreover, it is enough to show the slightly weaker statement by which there exists some finite measure $\bar{\mu}_0$ such that $\mu_N \xrightarrow{w} \bar{\mu}_0$, since then $\bar{\mu}_0$ must coincide with μ_0 because of the relations $G_N \xrightarrow{v} G_0$ and $K_N \to K_0$ uniformly in all bounded regions of $R^{k\nu}$, and K_0 is a continuous function. This implies that $\mu_N \xrightarrow{v} \mu_0$, and $\mu_0 = \bar{\mu}_0$.

There is a well-known theorem in probability theory about the equivalence between weak convergence of finite measures and the convergence of their Fourier transforms. It would be natural to apply this theorem for proving $\mu_N \xrightarrow{w} \bar{\mu}_0$. On the other hand, we have the additional information that the measures μ_N, $N = 1, 2, \dots$, are concentrated in the cubes $[-N\pi, N\pi)^{k\nu}$, since the spectral measure G is concentrated in $[-\pi, \pi)^\nu$. It is more fruitful to apply a version of the above mentioned theorem, where we can exploit our additional information. We formulate the following

Lemma 8.4. *Let $\mu_1, \mu_2, \dots$ be a sequence of finite measures on R^l such that $\mu_N(R^l \setminus [-C_N\pi, C_N\pi)^l) = 0$ for all $N = 1, 2, \dots$, with some sequence $C_N \to \infty$ as $N \to \infty$. Define the modified Fourier transform*

$$\varphi_N(t) = \int_{R^l} \exp\left\{i\left(\frac{[tC_N]}{C_N}, x\right)\right\} \mu_N(dx), \quad t \in R^l,$$

where $[tC_N]$ is the integer part of the vector $tC_N \in R^l$. (For an $x \in R^l$ its integer part $[x]$ is the vector $n \in \mathbb{Z}_l$ for which $x^{(p)} - 1 < n^{(p)} \le x^{(p)}$ if $x^{(p)} \ge 0$, and $x^{(p)} \le n^{(p)} < x^{(p)} + 1$ if $x^{(p)} < 0$ for all $p = 1, 2, \dots, l$.) If for all $t \in R^l$ the sequence $\varphi_N(t)$ tends to a function $\varphi(t)$ continuous at the origin, then the measures μ_N weakly tend to a finite measure μ_0, and $\varphi(t)$ is the Fourier transform of μ_0.

I make some comments on the conditions of Lemma 8.4. Let us observe that if the measures μ_N or a part of them are shifted with a vector $2\pi C_N u$ with some $u \in \mathbb{Z}_l$, then their modified Fourier transforms $\varphi_N(t)$ do not change because of the periodicity of the trigonometrical functions $e^{i(j/C_N,x)}$, $j \in \mathbb{Z}_l$. On the other hand, these new measures which are not concentrated in $[-C_N\pi, C_N\pi)^l$, have no limit. Lemma 8.4 states that if the measures μ_N are concentrated in the cubes $[-C_N\pi, C_N\pi)^l$, then the convergence of their modified Fourier transforms defined in Lemma 8.4, which is a weaker condition, than the convergence of their Fourier transforms, also implies their convergence to a limit measure.

Proof of Lemma 8.4. The proof is a natural modification of the proof about the equivalence of weak convergence of measures and the convergence of their Fourier transforms. First we show that for all $\varepsilon > 0$ there exits some $K = K(\varepsilon)$ such that

$$\mu_N(x\colon\ x \in R^l,\ |x^{(1)}| > K) < \varepsilon \quad \text{for all } N \geq 1. \tag{8.16}$$

As $\varphi(t)$ is continuous at the origin there is some $\delta > 0$ such that

$$|\varphi(0,\dots,0) - \varphi(t,0,\dots,0)| < \frac{\varepsilon}{2} \quad \text{if } |t| < \delta. \tag{8.17}$$

We have

$$0 \leq \mathrm{Re}\,[\varphi_N(0,\dots,0) - \varphi_N(t,0,\dots,0)] \leq 2\varphi_N(0,\dots,0) \tag{8.18}$$

for all $N = 1, 2, \dots$. The sequence in the middle term of (8.18) tends to

$$\mathrm{Re}\,[\varphi(0,\dots,0) - \varphi(t,0,\dots,0)]$$

as $N \to \infty$. The right-hand side of (8.18) is a bounded sequence, since it is convergent. Hence the dominated convergence theorem can be applied for the functions $\mathrm{Re}\,[\varphi_N(0,\dots,0) - \varphi_N(t,0,\dots,0)]$. Then we get because of the condition $C_N \to \infty$ and relation (8.17) that

$$\begin{aligned}
&\lim_{N\to\infty} \int_0^{[\delta C_N]/C_N} \frac{1}{\delta}\,\mathrm{Re}\,[\varphi_N(0,\dots,0) - \varphi_N(t,0,\dots,0)]\,dt \\
&\quad = \int_0^{\delta} \frac{1}{\delta}\,\mathrm{Re}\,[\varphi(0,\dots,0) - \varphi(t,0,\dots,0)]\,dt < \frac{\varepsilon}{2}
\end{aligned}$$

with the number $\delta > 0$ appearing in (8.17). Hence

$$\begin{aligned}
\frac{\varepsilon}{2} &> \lim_{N\to\infty} \int_0^{[\delta C_N]/C_N} \frac{1}{\delta}\,\mathrm{Re}\,[\varphi_N(0,\dots,0) - \varphi_N(t,0,\dots,0)]\,dt \\
&= \lim_{N\to\infty} \int \left(\frac{1}{\delta} \int_0^{[\delta C_N]/C_N} \mathrm{Re}\,[1 - e^{i[tC_N]x^{(1)}/C_N}]\,dt \right) \mu_N(\,dx) \\
&= \lim_{N\to\infty} \int \frac{1}{\delta C_N} \sum_{j=0}^{[\delta C_N]-1} \mathrm{Re}\left[1 - e^{ijx^{(1)}/C_N}\right] \mu_N(\,dx)
\end{aligned}$$

$$\geq \limsup_{N\to\infty} \int_{\{|x^{(1)}|>K\}} \frac{1}{\delta C_N} \sum_{j=0}^{[\delta C_N]-1} \text{Re}, \left[1 - e^{ijx^{(1)}/C_N}\right] \mu_N(dx)$$

$$= \limsup_{N\to\infty} \int_{\{|x^{(1)}|>K\}} \left(1 - \frac{1}{\delta C_N} \text{Re}\, \frac{1 - e^{i[\delta C_N]x^{(1)}/C_N}}{1 - e^{ix^{(1)}/C_N}}\right) \mu_N(dx)$$

with an arbitrary $K > 0$. (In the last but one step of this calculation we have exploited that $\frac{1}{\delta C_N} \sum\limits_{j=0}^{[\delta C_N]-1} \text{Re}\,[1 - e^{ijx^{(1)}/C_N}] \geq 0$ for all $x^{(1)} \in R^1$.)

Since the measure μ_N is concentrated in $\{x\colon x \in R^l,\ |x^{(1)}| \leq C_N \pi\}$, and

$$\text{Re}\, \frac{1 - e^{i[\delta C_N]x^{(1)}/C_N}}{1 - e^{ix^{(1)}/C_N}} = \frac{\text{Re}\, \left(i e^{-ix^{(1)}/2C_N} \left(1 - e^{i[\delta C_N]x^{(1)}/C_N}\right)\right)}{i(e^{-ix^{(1)}/2CN} - e^{ix^{(1)}/2C_N})}$$

$$\leq \frac{1}{\left|\sin\left(\dfrac{x^{(1)}}{2C_N}\right)\right|} \leq \frac{C_N \pi}{|x^{(1)}|}$$

if $|x^{(1)}| \leq C_N \pi$, (here we exploit that $|\sin u| \geq \frac{2}{\pi}|u|$ if $|u| \leq \frac{\pi}{2}$), hence we have with the choice $K = \frac{2\pi}{\delta}$

$$\frac{\varepsilon}{2} > \limsup_{N\to\infty} \int_{\{|x^{(1)}|>K\}} \left(1 - \left|\frac{\pi}{\delta x^{(1)}}\right|\right) \mu_N(dx) \geq \limsup_{N\to\infty} \frac{1}{2} \mu_N(|x^{(1)}| > K).$$

As the measures μ_N are finite the inequality $\mu_N(|x^{(1)}| > K) < \varepsilon$ holds for each index N with a constant $K = K(N)$ that may depend on N. Hence the above inequality implies that formula (8.16) holds for all $N \geq 1$ with a possibly larger index K that does not depend on N.

Applying the same argument to the other coordinates we find that for all $\varepsilon > 0$ there exists some $C(\varepsilon) < \infty$ such that

$$\mu_N\left(R^l \setminus [-C(\varepsilon), C(\varepsilon)]^l\right) < \varepsilon \quad \text{for all } N = 1, 2, \ldots.$$

Consider the usual Fourier transforms

$$\tilde{\varphi}_N(t) = \int_{R^l} e^{i(t,x)} \mu_N(dx), \quad t \in R^l.$$

Then

$$|\varphi_N(t) - \tilde{\varphi}_N(t)| \leq 2\varepsilon + \int_{[-C(\varepsilon),C(\varepsilon)]} \left|e^{i(t,x)} - e^{i([tC_N]/C_N,x)}\right| \mu_N(dx)$$

$$\leq 2\varepsilon + \frac{lC(\varepsilon)}{C_N} \mu_N(R^l)$$

for all $\varepsilon > 0$. Hence $\tilde{\varphi}_N(t) - \varphi_N(t) \to 0$ as $N \to \infty$, and $\tilde{\varphi}_N(t) \to \varphi(t)$. (Observe that $\mu_N(R^l) = \varphi_N(0) \to \varphi(0) < \infty$ as $N \to \infty$, hence the measures $\mu_N(R^l)$ are uniformly bounded, and $C_N \to \infty$ by the conditions of Lemma 8.4.) Then Lemma 8.4 follows from standard theorems on Fourier transforms. □

We return to the proof of Theorem 8.2. We apply Lemma 8.4 with $C_N = N$ and $l = k\nu$ for the measures μ_N defined in (8.15). Because of the middle term in (8.14) we can write the modified Fourier transform φ_N of the measure μ_N as

$$\varphi_N(t_1, \dots, t_k) = \sum_{r=1}^{p} \sum_{s=1}^{p} c_r c_s \psi_N(t_1 + n_r - n_s, \dots, t_k + n_r - n_s) \tag{8.19}$$

with

$$\begin{aligned} \psi_N(t_1, \dots, t_r) &= \frac{1}{N^{2\nu}} \int \exp\left\{ i \frac{1}{N}((j_1, x_1) + \dots + (j_k, x_k)) \right\} \\ &\qquad \sum_{u \in B_0^N} \sum_{v \in B_0^N} \exp\left\{ i \left(\frac{u - v}{N}, x_1 + \dots + x_k \right) \right\} G_N(dx_1) \dots G_N(dx_k) \\ &= \frac{1}{N^{2\nu - k\alpha} L(N)^k} \sum_{u \in B_0^N} \sum_{v \in B_0^N} r(u - v + j_1) \cdots r(u - v + j_k), \end{aligned} \tag{8.20}$$

where $j_p = [t_p N]$, $t_p \in R^\nu$, $p = 1, \dots, k$.

The asymptotical behaviour of $\psi_N(t_1, \dots, t_k)$ for $N \to \infty$ can be investigated by the help of the last relation and formula (8.1). Rewriting the last double sum in the form of a single sum by fixing first the variable $l = u - v \in [-N, N]^\nu \cap \mathbb{Z}_\nu$, and then summing up for l one gets

$$\psi_N(t_1, \dots, t_k) = \int_{[-1,1]^\nu} f_N(t_1, \dots, t_k, x)\, dx$$

with

$$\begin{aligned} &f_N(t_1, \dots, t_k, x) \\ &\quad = \left(1 - \frac{[|x^{(1)} N|]}{N}\right) \cdots \left(1 - \frac{[|x^{(\nu)} N|]}{N}\right) \frac{r([xN] + j_1)}{N^{-\alpha} L(N)} \cdots \frac{r([xN] + j_k)}{N^{-\alpha} L(N)}. \end{aligned}$$

(In the above calculation we exploited that in the last sum of formula (8.20) the number of pairs (u, v) for which $u - v = l = (l_1, \dots, l_\nu)$ equals $(N - |l_1|) \cdots (N - |l_\nu|)$.)

Let us fix some vector $(t_1, \dots, t_k) \in R^{k\nu}$. It can be seen with the help of formula (8.1) that for all $\varepsilon > 0$ the convergence

$$f_N(t_1, \dots, t_k, x) \to f_0(t_1, \dots, t_k, x) \tag{8.21}$$

holds uniformly with the limit function

$$f_0(t_1,\dots,t_k,x)=(1-|x^{(1)}|)\dots(1-|x^{(\nu)}|)\frac{a\left(\frac{x+t_1}{|x+t_1|}\right)}{|x+t_1|^\alpha}\cdots\frac{a\left(\frac{x+t_k}{|x+t_k|}\right)}{|x+t_k|^\alpha} \tag{8.22}$$

on the set $x\in[-1,1]^\nu\setminus\bigcup\limits_{p=1}^{k}\{x\colon |x+t_p|>\varepsilon\}$.

We claim that

$$\psi_N(t_1,\dots,t_k)\to\psi_0(t_1,\dots,t_k)=\int_{[-1,1]^\nu} f_0(t_1,\dots,t_k,x)\,dx,$$

and ψ_0 is a continuous function.

This relation implies that $\mu_N\xrightarrow{w}\mu_0$. To prove it, it is enough to show beside formula (8.21) that

$$\left|\int_{|x+t_p|<\varepsilon} f_0(t_1,\dots,t_k,x)\,dx\right|<C(\varepsilon),\quad p=1,\dots,k, \tag{8.23}$$

and

$$\int_{|x+t_p|<\varepsilon} |f_N(t_1,\dots,t_k,x)|\,dx<C(\varepsilon),\quad p=1,\dots,k,\quad\text{and } N=1,2,\dots \tag{8.24}$$

with a constant $C(\varepsilon)$ such that $C(\varepsilon)\to0$ as $\varepsilon\to0$.

By formula (8.22) and Hölder's inequality

$$\left|\int_{|x+t_p|<\varepsilon} f_0(t_1,\dots,t_k,x)\,dx\right|\le C\prod_{1\le l\le k,\,l\ne p}\left[\int_{x\in[-1,1]^\nu}|x+t_l|^{-k\alpha}\,dx\right]^{1/k}$$
$$\left[\int_{|x+t_p|\le\varepsilon}|x+t_p|^{-k\alpha}\,dx\right]^{1/k}\le C'\varepsilon^{\nu/k-\alpha}$$

with some appropriate $C>0$ and $C'>0$, since $\nu-k\alpha>0$, and $a(\cdot)$ is a bounded function. Similarly,

$$\int_{|x+t_p|<\varepsilon}|f_N(t_1,\dots,t_k,x)|\,dx\le\prod_{1\le l\le k,\,l\ne p}\left[\int_{x\in[-1,1]^\nu}\frac{|r([xN]+j_l)|^k}{N^{-k\alpha}L(N)^k}\,dx\right]^{1/k},$$
$$\left[\int_{|x+t_p|\le\varepsilon}\frac{|r([xN]+j_p)|^k}{N^{-k\alpha}L(N)^k}\,dx\right]^{1/k}. \tag{8.25}$$

It is not difficult to see with the help of Karamata's theorem that if $L(t)$, $t \ge 1$, is a slowly varying function which is bounded in all finite intervals, then for all numbers $\eta > 0$ and $K > 0$ there are some constants $K_1 = K_1(\eta, K) > 0$, and $C = C(\eta, K) > 0$ together with a threshold index $N_0 = N_0(\eta, K)$ such that

$$\frac{L(uN)}{L(N)} \le C u^{-\eta} \quad \text{if } uN > K_1, \quad u \le K, \quad \text{and } N \ge N_0.$$

Hence formula (8.1) implies that

$$|r([xN] + [t_{lN}]) = |r([xN] + j_l)| \le C N^{-\alpha} L(N) |x + t_l|^{-\alpha-\eta}$$
$$\text{if } |x + t_l| \le K \text{ and } N \ge N_0. \tag{8.26}$$

Relation (8.26) follows from the previous relation and (8.1) if $|[xN] + [t_{lN}]| \ge K_1$. It also holds if $|[xN] + [t_{lN}]| \le K_1$, since in this case the left-hand side can be bounded by the inequality $|r([xN] + [t_{lN}]| \le 1$, while the right-hand side of (8.26) is greater than 1 with the choice of a sufficiently large constant C (depending on η and K_1). This follows from the relation $|x + t|^{-\alpha-\eta} = N^{\alpha+\eta} |N(x+t)|^{-\alpha-\eta} \ge C_1 N^{\alpha+\eta}$ if $|[xN] + [t_{lN}]| \le K_1$, and $L(N) \ge N^{-\eta}$.

We get with the help of (8.26) that

$$\int_{|x+t_p|<\varepsilon} \frac{|r([xN] + j_p)|^k}{N^{-k\alpha} L(N)^k} \, dx \le B \int_{|x+t_p|<\varepsilon} |x + t_p|^{-k(\alpha+\eta)} \, dx \le B' \varepsilon^{\nu - k(\alpha+\eta)}$$

$$\int_{x \in [-1,1]^\nu} \frac{|r([xN] + j_l)|^k}{N^{-k\alpha} L(N)^k} \, dx \le B''.$$

for a sufficiently small constant $\eta > 0$ with some constants $B, B', B'' < \infty$ depending on η and t_p, $1 \le p \le k$.

Therefore we get from (8.25), by choosing an $\eta > 0$ such that $k(\alpha + \eta) < \nu$, that the inequality

$$\int_{|x+t_p|<\varepsilon} |f_N(t_1, \dots, t_k, x)| \, dx \le C \varepsilon^{\nu/k - (\alpha+\eta)}$$

holds with some $C < \infty$. The right-hand side of this inequality tends to zero as $\varepsilon \to 0$. Hence we proved beside (8.21) formulae (8.23) and (8.24), and they have the consequence that $\psi_N(t_1, \dots, t_k) \to \psi_0(t_1, \dots, t_k)$. Since $\psi(t_1, \dots, t_k)$ is a continuous function relation (8.19) with Lemma 8.4 imply that the measures μ_N introduced in (8.18) converge weakly to a probability measure as $N \to \infty$, and as we saw at the beginning of the proof of Theorem 8.2 this limit measure must be μ_0.

Hence we can apply Lemma 8.3 for the spectral measures G_N and functions $K_N(\cdot)$, $N = 0, 1, 2, \dots$, defined in Theorem 8.2. In this application of Lemma 8.3 we choose no rectangles P_N. The convergence $G_N \xrightarrow{v} G_0$ follows from Lemma 8.1.

Conditions (a) and (b) also hold with the choice of a sufficiently large number $A = A(\varepsilon)$. The hard point of the proof was the checking of condition (b). This follows from the relation $\mu_N \xrightarrow{w} \mu_0$. Thus we have proved Theorem 8.2 with the help of Lemma 8.1. □

It remained to prove Lemma 8.1.

Proof of Lemma 8.1. Introduce the notation

$$K_N(x) = \prod_{j=1}^{\nu} \frac{e^{ix^{(j)}} - 1}{N(e^{ix^{(j)}/N} - 1)}, \qquad N = 1, 2, \dots,$$

and

$$K_0(x) = \prod_{j=1}^{\nu} \frac{e^{ix^{(j)}} - 1}{ix^{(j)}}.$$

Let us consider the measures μ_N defined in formula (8.15) in the special case $k = 1$ with $p = 1$, $c_1 = 1$ in the definition of the function $K_N(\cdot)$, i.e. put

$$\mu_N(A) = \int_A |K_N(x)|^2 \, G_N(dx), \quad A \in \mathscr{B}^\nu, \quad N = 1, 2, \dots.$$

We have already seen in the proof of Theorem 8.2 that $\mu_N \xrightarrow{w} \mu_0$ with some finite measure μ_0, and the Fourier transform of μ_0 is

$$\varphi_0(t) = \int_{[-1,1]^\nu} (1 - |x^{(1)}|) \cdots (1 - |x^{(\nu)}|) \frac{a\left(\frac{x+t}{|x+t|}\right)}{|x+t|^\alpha} \, dx.$$

Moreover, since $|K_N(x)|^2 \to |K_0(x)|^2$ uniformly in any bounded domain, it is natural to expect that $G_N \xrightarrow{v} G_0$ with $G_0(dx) = \frac{1}{|K_0(x)|^2} \mu_0(dx)$. But $K_0(x) = 0$ in some points, and the function $K_0(\cdot)^{-2}$ is not continuous in these points. As a consequence, we cannot give a direct proof of the above statement. Hence we apply instead a modified version of this method. First we prove the following result about the behaviour of the restrictions of the measures G_N to appropriate cubes:

For all $T \ge 1$ there is a finite measure G_0^T concentrated on $(-T\pi, T\pi)^\nu$ such that

$$\lim_{N \to \infty} \int f(x) \, G_N(dx) = \int f(x) \, G_0^T(dx) \tag{8.27}$$

for all continuous functions f which vanish outside the cube $(-T\pi, T\pi)^\nu$.

Indeed, let a continuous function f vanish outside the cube $(-T\pi, T\pi)^\nu$ with some $T \geq 1$. Put $M = [\frac{N}{2T}]$. Then

$$\begin{aligned}\int f(x)G_N(dx) &= \frac{N^\alpha}{L(N)} \cdot \frac{L(M)}{M^\alpha} \int f\left(\frac{N}{M}x\right) G_M(dx) \\ &= \frac{N^\alpha L(M)}{M^\alpha L(N)} \int f\left(\frac{N}{M}x\right) |K_M(x)|^{-2} \mu_M(dx) \\ &\to (2T)^\alpha \int f(2Tx)|K_0(x)|^{-2}\mu_0(dx) \\ &= \int f(x) \frac{(2T)^\alpha}{|K_0(\frac{x}{2T})|^2} \mu_0\left(\frac{dx}{2T}\right) \quad \text{as } N \to \infty,\end{aligned}$$

because $f(\frac{N}{M}x)|K_M(x)|^{-2}$ vanishes outside the cube $[-\pi, \pi]^\nu$, the limit relation

$$f(\frac{N}{M}x)|K_M(x)|^{-2} \to f(2Tx)|K_0(x)|^{-2}$$

holds uniformly, (the function $K_0(\cdot)^{-2}$ is continuous in the cube $[-\pi, \pi]^\nu$), and $\mu_M \xrightarrow{w} \mu_0$ as $N \to \infty$. Hence relation (8.27) holds if we define G_0^T as the restriction of the measure $\frac{(2T)^\alpha}{|K_0(\frac{x}{2T})|^2} \mu_0\left(\frac{dx}{2T}\right)$ to the cube $(-T\pi, T\pi)^\nu$. The measures G_0^T appearing in (8.27) are consistent for different parameters T, i.e. G_0^T is the restriction of the measure $G_0^{T'}$ to the cube $(-T\pi, T\pi)^\nu$ if $T' > T$. This follows from the fact that $\int f(x)G_0^T(dx) = \int f(x)G_0^{T'}(dx)$ for all continuous functions with support in $(-T, T)^\nu$. We claim that by defining the measure G_0 by the relation $G_0(A) = G_0^T(A)$ for a bounded set A and such number $T > 1$ for which $A \subset (-T\pi, T\pi)^\nu$ we get such a locally finite measure G_0 for which $G_N \xrightarrow{v} G_0$. The above mentioned vague convergence is a direct consequence of (8.27) and the definition of G_0, but to give a complete proof we have to show that G_0 is really a (σ-additive) measure.

Actually it is enough to prove that the restriction of G_0 to the bounded, measurable sets is σ-additive, because it follows then from standard results in measure theory that it has a unique σ-additive extension to $\mathscr{B}^\nu$. But this is an almost direct consequence of the definition of G_0. The desired σ-additivity clearly holds, since if $A = \bigcup\limits_{n=1}^{\infty} A_n$, the set A is bounded, and the sets A_n, $n = 1, 2, \dots$, are disjoint, then there is a number $T > 1$ such that $A \subset (-T\pi, T\pi)^\nu$, the same relation holds for the sets A_n, and the σ-additivity of G_0^T implies that $G_0(A) = \sum\limits_{n=1}^{\infty} G_0(A_n)$.

As $G_N \xrightarrow{v} G_0$, and $|K_N(x)|^2 \to |K_0(x)|^2$ uniformly in all bounded regions, the relation $\mu_N \xrightarrow{v} \bar{\mu}_0$ holds with the measure $\bar{\mu}_0$ defined as $\bar{\mu}_0(A) = \int_A |K_0(x)|^2 G_0(dx)$, $A \in \mathscr{B}^\nu$. Since $\mu_N \xrightarrow{w} \mu_0$ the measures μ_0 and $\bar{\mu}_0$ must coincide, i.e.

$$\mu_0(A) = \int_A |K_0(x)|^2 \, G_0(\,dx), \quad A \in \mathscr{B}^\nu.$$

Relation (8.4) expresses the fact that φ_0 is the Fourier transform of μ_0.

It remained to prove the homogeneity property (8.3) of the measure G_0. For this goal let us extend the definition of the measures G_N given in (8.2) to all non-negative real numbers u. It is easy to see that the relation $G_u \xrightarrow{v} G_0$ as $u \to \infty$ remains valid. Hence we get for all fixed $s > 0$ and continuous functions f with compact support that

$$\begin{aligned}\int f(x)G_0(\,dx) &= \lim_{u\to\infty} \int f(x)\,G_u(\,dx) = \lim_{u\to\infty} \frac{s^\alpha L(\frac{u}{s})}{L(u)} \int f(sx)G_{\frac{u}{s}}(\,dx)\\ &= s^\alpha \int f(sx)G_0(\,dx) = \int f(x)s^\alpha G_0\left(\frac{dx}{s}\right).\end{aligned}$$

This identity implies the homogeneity property (8.3) of G_0. Lemma 8.1 is proved. □

The next result is a generalization of Theorem 8.2.

Theorem 8.2′. *Let X_n, $n \in \mathbb{Z}_\nu$, be a stationary Gaussian field with a correlation function $r(n)$ defined in (8.1). Let $H(x)$ be a real function with the properties $EH(X_n) = 0$ and $EH(X_n)^2 < \infty$. Let us consider the orthogonal expansion*

$$H(x) = \sum_{j=1}^{\infty} c_j H_j(x), \quad \sum c_j^2 j! < \infty, \tag{8.28}$$

of the function $H(\cdot)$ by the Hermite polynomials H_j (with leading coefficients 1). Let k be the smallest index in this expansion such that $c_k \neq 0$. If $0 < k\alpha < \nu$ for the parameter α in (8.1), and the field Z_n^N is defined by the field $\xi_n = H(X_n)$, $n \in \mathbb{Z}_\nu$, and formula (1.1), then the multi-dimensional distributions of the fields Z_n^N with $A_N = N^{\nu-k\alpha/2}L(N)^{k/2}$ tend to those of the fields $c_k Z_n^*$, $n \in \mathbb{Z}_\nu$, where the field Z_n^* is the same as in Theorem 8.2.

Proof of Theorem 8.2′. Define $H'(x) = \sum\limits_{j=k+1}^{\infty} c_j H_j(x)$ and $Y_n^N = \frac{1}{A_N} \sum\limits_{l \in B_n^N} H'(X_l)$. Because of Theorem 8.2 in order to prove Theorem 8.2′ it is enough to show that

$$E(Y_n^N)^2 \to 0 \quad \text{as } N \to \infty.$$

It follows from Corollary 5.5 that $EH_j(X_n)H_l(X_m) = \delta_{j,l}\, j!(EX_nX_m)^j = \delta_{j,l}\, j!r(n-m)^j$, where $\delta_{j,l} = 0$ if $j \neq l$, and $\delta_{j,l} = 1$ if $j = l$. Hence

$$E(Y_n^N)^2 = \frac{1}{A_N^2} \sum_{j=k+1}^{\infty} c_j^2 j! \sum_{s,t \in B_n^N} [r(s-t)]^j.$$

Some calculation yields with the help of this identity and formula (8.1) that

$$E(Y_n^N)^2 = \frac{1}{A_N^2}\left[O(N^{2\nu-(k+1)\alpha}L(N)^{k+1}) + O(N^\nu)\right] \to 0.$$

(Observe that we imposed the condition $\sum c_j^2 j! < \infty$ which is equivalent to the condition $EH(X_n)^2 < \infty$.) Theorem 8.2′ is proved. □

Let us consider a slightly more general version of the problem investigated in Theorem 8.2′. Take a stationary Gaussian random field X_n, $EX_n = 0$, $EX_n^2 = 1$, $n \in \mathbb{Z}_\nu$ with a correlation function satisfying relation (8.1), and the field $\xi_n = H(X_n)$, $n \in \mathbb{Z}_\nu$, subordinated to it with a general function $H(x)$ such that $EH(X_n) = 0$ and $EH(X_n)^2 < \infty$. We are interested in the large-scale limit of such random fields. Take the Hermite expansion (8.28) of the function $H(x)$, and let k be the smallest such index for which $c_k \neq 0$ in the expansion (8.28). In Theorem 8.2′ we solved this problem if $0 < k\alpha < \nu$. We are interested in the question what happens in the opposite case when $k\alpha > \nu$. Let me remark that in the case $k\alpha \geq \nu$ the field Z_n^*, $n \in \mathbb{Z}_\nu$, which appeared in the limit in Theorem 8.2′ does not exist. The Wiener–Itô integral defining Z_n^* is meaningless, because the integral which should be finite to guarantee the existence of the Wiener–Itô integral is divergent in this case. Next I formulate a general result which contains the answer to the above question as a special case.

Theorem 8.5. *Let us consider a stationary Gaussian random field X_n, $EX_n = 0$, $EX_n^2 = 1$, $n \in \mathbb{Z}_n$, with correlation function $r(n) = EX_m X_{m+n}$, $m, n \in \mathbb{Z}_\nu$. Take a function $H(x)$ on the real line such that $EH(X_n) = 0$ and $EH(X_n)^2 < \infty$. Take the Hermite expansion (8.28) of the function $H(x)$, and let k be smallest index in this expansion such that $c_k \neq 0$. If*

$$\sum_{n \in \mathbb{Z}_\nu} |r(n)|^k < \infty, \tag{8.29}$$

then the limit

$$\lim_{N\to\infty} EZ_n^N(H_l)^2 = \lim_{N\to\infty} N^{-\nu} \sum_{i \in B_n^N} \sum_{j \in B_n^N} r^l(i-j) = \sigma_l^2 l!$$

exists for all indices $l \geq k$, where $Z_n^N(H_l)$ is defined in (1.1) with $A_N = N^{\nu/2}$, and $\xi_n = H_l(X_n)$ with the l-th Hermite polynomial $H_l(x)$ with leading coefficient 1. Moreover, also the inequality

$$\sigma^2 = \sum_{l=k}^{\infty} c_l^2 l! \sigma_l^2 < \infty$$

holds.

The finite dimensional distributions of the random field $Z_n^N(H)$ defined in (1.1) with $A_N = N^{\nu/2}$ and $\xi_n = H(X_n)$ tend to the finite dimensional distributions of a random field σZ_n^* with the number σ defined in the previous relation, where Z_n^*, $n \in \mathbb{Z}_\nu$, are independent, standard normal random variables.

Theorem 8.5 can be applied if the conditions of Theorem 8.2′ hold with the only modification that the condition $k\alpha < \nu$ is replaced by the relation $k\alpha > \nu$. In this case the relation (8.29) holds, and the large-scale limit of the random field Z_n^N, $n \in \mathbb{Z}_\nu$ with normalization $A_N = N^{\nu/2}$ is a random field consisting of independent standard normal random variables multiplied with the number σ. There is a slight generalization of Theorem 8.5 which also covers the case $k\alpha = \nu$. In this result we assume instead of the condition (8.29) that $\sum\limits_{n \in \bar{B}_N} r(n)^k = L(N)$ with a slowly varying function $L(\cdot)$, where $\bar{B}_N = \{(n_1, \dots, n_\nu) \in \mathbb{Z}_\nu\colon -N \le n_j \le N,\ 1 \le j \le \nu\}$, and some additional condition is imposed which states that an appropriately defined finite number $\sigma^2 = \lim\limits_{N \to \infty} \sigma_N^2$, which plays the role of the variance of the random variables in the limiting field, exists. There is a similar large scale limit in this case as in Theorem 8.5, the only difference is that the norming constant in this case is $A_N = N^{\nu/2} L(N)^{1/2}$. This result has the consequence that if the conditions of Theorem 8.2′ hold with the only difference that $k\alpha = \nu$ instead of $k\alpha < \nu$, then the large scale limit exists with norming constants $A_N = N^{\nu/2} L(N)$ with an appropriate slowly varying function $L(\cdot)$, and it consists of independent Gaussian random variables with expectation zero.

The proof of Theorem 8.5 and its generalization that we did not formulate here explicitly appeared in paper [3]. I omit its proof, I only make some short explanation about it.

In the proof we show that all moments of the random variables Z_n^N converge to the corresponding moments of the random variables Z_n^* as $N \to \infty$. The moments of the random variables Z_n^N can be calculated by means of the diagram formula if we either rewrite them in the form of a Wiener–Itô integral or apply a version of the diagram formula which gives the moments of Wick polynomials instead of Wiener–Itô integrals. In both cases the moments can be expressed explicitly by means of the correlation function of the underlying Gaussian random field. The most important step of the proof is to show that we can select a special subclass of (closed) diagrams, called regular diagrams in [3] which yield the main contribution to the moment $E(Z_n^N)^M$, and their contribution can be simply calculated. The contribution of all remaining diagrams is $o(1)$, hence it is negligible. For the sake of simplicity let us restrict our attention to the case $H(x) = H_k(x)$, and let us explain the definition of the regular diagrams in this special case.

If M is an even number, then take the partitions $\{k_1, k_2\}$, $\{k_3, k_4\}, \dots$, $\{k_{M-1}, k_M\}$ of the set $\{1, \dots, M\}$ to subsets consisting of exactly two elements, to define the regular diagrams. They are those (closed) diagrams for which we can choose one of the above partitions in such a way that the diagram contains only edges connecting vertices from the k_{2j-1}-th and k_{2j}-th row with some $1 \le j \le \frac{M}{2}$, where $\{k_{2j-1}, k_{2j}\}$ is an element of the partition we have chosen. If M is an odd number, then there is no regular diagram.

In Theorems 8.2 and 8.2′ we investigated some very special subordinated fields. The next result shows that the same limiting field as the one in Theorem 8.2 appears in a much more general situation.

Let us define the field

$$\xi_n = \sum_{j=k}^{\infty} \frac{1}{j!} \int e^{i(n,x_1+\cdots+x_j)} \alpha_j(x_1,\dots,x_j)\, Z_G(dx_1)\dots Z_G(dx_j), \quad n \in \mathbb{Z}_\nu, \tag{8.30}$$

where Z_G is the random spectral measure adapted to a Gaussian field X_n, $n \in \mathbb{Z}_\nu$, with correlation function satisfying (8.1) with $0 < \alpha < \frac{\nu}{k}$.

Theorem 8.6. *Let the fields Z_n^N be defined by formulae (8.30) and (8.1) with $A_N = N^{\nu-k\alpha/2}L(N)^{k/2}$. The multi-dimensional distributions of the fields Z_n^N tend to those of the field $\alpha_k(0,\dots,0)Z_n^*$ where the field Z_n^* is the same as in Theorem 8.2 if the following conditions are fulfilled:*

(i) $\alpha_k(x_1,\dots,x_k)$ is a bounded function, continuous at the origin, and such that $\alpha_k(0,\dots,0) \neq 0$.

(ii) $$\sum_{j=k=1}^{\infty} \frac{1}{j!} \frac{N^{-(j-k)\alpha}}{L(N)^{j-k}} \int_{R^{j\nu}} \left|\alpha_j\left(\frac{x_1}{N},\dots,\frac{x_j}{N}\right)\right|^2 \frac{1}{N^{2\nu}} \left|\sum_{j\in B_0^N} e^{i(l/N,x_1+\cdots+x_j)}\right|^2 G_N(dx_1)\dots G_N(dx_j) \to 0,$$

where G_N is defined in (8.2).

Proof of Theorem 8.6. The proof is very similar to those of Theorems 8.2 and 8.2′. The same argument as in the proof of Theorem 8.2′ shows that because of condition (ii) ξ_n can be substituted in the present proof by the following expression:

$$\xi_n' = \frac{1}{k!} \int e^{i(n,x_1+\cdots+x_k)} \alpha_k(x_1,\dots,x_k) Z_G(dx_1)\dots Z_G(dx_k), \quad n \in \mathbb{Z}_\nu.$$

Then a natural modification in the proof of Theorem 8.2 implies Theorem 8.6. The main point in this modification is that we have to substitute the measures μ_N defined in formula (8.15) by the following measure $\bar\mu_N$:

$$\bar\mu_N(A) = \int_A |K_N(x_1,\dots,x_k)|^2 \left|\alpha_k\left(\frac{x_1}{N},\dots,\frac{x_k}{N}\right)\right|^2 G_N(dx_1)\dots G_N(dx_k), \quad A \in \mathcal{B}^{k\nu},$$

and to observe that because of condition (i) the limit relation $\mu_N \xrightarrow{w} \mu_0$ implies that $\bar\mu_N \xrightarrow{w} |\alpha_k(0,\dots,0)|^2\mu_0$. □

The main problem in applying Theorem 8.6 is to check conditions (i) and (ii). We remark without proof that any field $\xi_n = H(X_{s_1+n}, \dots, X_{s_p+n}), s_1, \dots, s_p \in \mathbb{Z}_\nu$ and $n \in \mathbb{Z}_\nu$, for which $E\xi_n^2 < \infty$ satisfies condition (ii). This is proved in Remark 6.2 of [9]. If the conditions (i) or (ii) are violated, then a limit of different type may appear.

Finally we quote such a result without proof. Actually the proof is similar to that of Theorem 8.2. At this point the general formulation of Lemma 8.3 is useful. (See [25] for a proof.) Here we restrict ourselves to the case $\nu = 1$. The limiting field appearing in this result belongs to the class of self-similar fields constructed in Remark 6.5.

Let a_n, $n = \dots, -1, 0, 1, \dots$, be a sequence of real numbers such that

$$\begin{aligned} a_n &= C(1)n^{-\beta-1} + o(n^{-\beta-1}) \quad \text{if } n \ge 0 \\ a_n &= C(2)|n|^{-\beta-1} + o(|n|^{-\beta-1}) \quad \text{if } n < 0 \end{aligned} \qquad -1 < \beta < 1. \tag{8.31}$$

Let X_n, $n = \dots, -1, 0, 1, \dots$, be a stationary Gaussian sequence with correlation function $r(n) = EX_0X_n = |n|^{-\alpha}L(|n|)$, $0 < \alpha < 1$, where $L(\cdot)$ is a slowly varying function. Define the field ξ_n, $n = \dots, -1, 0, 1, \dots$, as

$$\xi_n = \sum_{m=-\infty}^{\infty} a_m H_k(X_{m+n}). \tag{8.32}$$

Theorem 8.7. *Let a sequence ξ_n, $n = \dots, -1, 0, 1, \dots$, be defined by (8.31) and (8.32). Let $0 < k\alpha < 1$, $0 < 1 - \beta - \frac{k}{2}\alpha < 1$, and let one of the following conditions be satisfied.*

(a) $0 < \beta < 1$, *and* $\sum_{n=-\infty}^{\infty} a_n = 0$.

(b) $0 > \beta > -1$.

(c) $\beta = 0$, $C(1) = -C(2)$, *and* $\sum_{n=0}^{\infty} |a_n + a_{-n}| < \infty$.

Let us define the sequences Z_n^N by formula (1.1) with $A_N = N^{1-\beta-k\alpha/2}L(N)^{k/2}$ and the above defined field ξ_n. The multi-dimensional distributions of the sequences Z_n^N tend to those of the sequences $D^{-k}Z_n^*(\alpha, \beta, a, b, c)$, where

$$Z_n^*(\alpha, \beta, k, b, c) = \int \tilde{\chi}_n(x_1 + \dots + x_k)$$
$$\left[b|x_1 + \dots + x_k|^\beta + ic|x_1 + \dots + x_k|^\beta \operatorname{sign}(x_1 + \dots + x_k)\right]$$
$$|x_1|^{(\alpha-1)/2} \dots |x_k|^{(\alpha-1)/2} W(dx_1) \dots W(dx_k),$$

$W(\cdot)$ denotes the white noise field, i.e. a random spectral measure corresponding to the Lebesgue measure, and the constants D, b and c are defined as $D = 2\Gamma(\alpha)\cos(\frac{\alpha}{2}\pi)$, *and*

$b = 2[C(1) + C(2)]\Gamma(-\beta)\sin(\frac{\beta+1}{2}\pi)$, *and* $c = 2[C(1) - C(2)]\Gamma(-\beta)\cos(\frac{\beta+1}{2}\pi)$ *in cases (a) and (b), and*

$b = \sum\limits_{n=-\infty}^{\infty} a_n$, *and* $c = C(1)$ *in case (c).*

Chapter 9
History of the Problems: Comments

Chapter 1

In statistical physics the problem formulated in this chapter appeared at the investigation of some physical models at critical temperature. A discussion of this problem and further references can be found in the fourth chapter of the forthcoming book of Ya.G. Sinai [34]. (Here and in the later part of Chap. 9 we did not change the text of the first edition. Thus expressions like forthcoming book, recent paper, etc. refer to the time when the first version of this Lecture Note appeared.) The first example of a limit theorem for partial sums of random variables which is considerably different from the independent case was given by M. Rosenblatt in [29]. Further results in this direction were proved by R.L. Dobrushin, H. Kesten and F. Spitzer, P. Major, M. Rosenblatt and M.S. Taqqu [7–9, 25, 30, 31, 35, 38]. In most of these papers only the one-dimensional case is considered, i.e. the case when $R^\nu = R^1$, and it is formulated in a different but equivalent way. In most of these works the joint distribution of the random variables $A_N^{-1} \sum\limits_{j=1}^{Nt]} \xi_j$, $0 < t < \infty$, was considered.

Similar problems also appeared in the theory of infinite particle systems. The large-scale limit of the so-called voter model and of infinite particle branching Brownian motions were investigated in papers [2, 6, 18, 24]. It was proved that in these models the limit is a Gaussian self-similar field with a non-typical normalization. The investigation of the large-scale limit would be very natural for many other infinite particle systems, but in most cases this problem is hopelessly difficult.

The notion of subordinated fields in the present context first appeared at Dobrushin [7]. It is natural to expect that there exists a large class of self-similar fields which cannot be obtained as subordinated fields. Nevertheless the present techniques are not powerful enough for finding them.

P. Major, *Multiple Wiener-Itô Integrals*, Lecture Notes in Mathematics 849, DOI 10.1007/978-3-319-02642-8_9,

The approach to the problem is different in statistical physics. In statistical physics one looks for self-similar fields which satisfy some conditions formulated in accordance to physical considerations. One tries to describe these fields with the help of a power series which is the Radon–Nykodim derivative of the field with respect to a Gaussian field. The deepest result in this direction is a recent paper of P.M. Bleher and M.D. Missarov [1] where the required formal power series is described. This result enables one to calculate several critical indices interesting for physicists, but the task of proving that this formal expression defines an existing field seems to be very hard. It is also an open problem whether the class of self-similar fields constructed via multiple Wiener–Itô integrals contains the non-Gaussian self-similar fields interesting for statistical physics. Some experts are very skeptical in this respect. The Gaussian self-similar fields are investigated in [7, 33]. A more thorough investigation is presented in [11].

The notion of generalized random fields was introduced by I.M. Gelfand. A detailed discussion can be found in the book [16], where the properties of Schwartz spaces we need can also be found.

In the definition of generalized fields the class of test functions $\mathcal{S}$ can be substituted by other linear topological spaces consisting of real valued functions. The most frequently considered space, beside the space $\mathcal{S}$, is the space $\mathcal{D}$ of infinitely many times differentiable functions with compact support. In paper [7] Dobrushin also considered the space $\mathcal{S}^r \subset \mathcal{S}$, which consists of the functions $\varphi \in \mathcal{S}$ satisfying the additional relation $\int x^{(1)^{j_1}} \cdots x^{(\nu)^{j_\nu}} \varphi(x)\,dx = 0$, provided that $j_1 + \cdots + j_\nu < r$. He considered this class of test functions, because there are much more continuous linear functionals over $\mathcal{S}^r$ than over $\mathcal{S}$, and this property of $\mathcal{S}^r$ can be exploited in certain investigations. Generally no problem arises in the proofs if the space of test functions $\mathcal{S}$ is substituted by $\mathcal{S}^r$ or $\mathcal{D}$ in the definition of generalized fields.

Two generalized fields $X(\varphi)$ and $\bar{X}(\varphi)$ can be identified if $X(\varphi) \stackrel{\Delta}{=} \bar{X}(\varphi)$ for all $\varphi \in \mathcal{S}$. Let me remark that this relation also implies that the multi-dimensional distributions of the random vectors $(X(\varphi_1), \dots, X(\varphi_n))$ and $(\bar{X}(\varphi_1), \dots, \bar{X}(\varphi_n))$ coincide for all $\varphi_1, \dots, \varphi_n \in \mathcal{S}$. As $\mathcal{S}$ is a linear space, this relation can be deduced from property (a) of generalized fields by exploiting that two distribution functions on R^n agree if and only if their characteristic functions agree.

Let $\mathcal{S}'$ denote the space of continuous linear functionals over $\mathcal{S}$, and let $\mathcal{A}_{\mathcal{S}'}$ be the σ-algebra over $\mathcal{S}'$ generated by the sets $A(\varphi, a) = \{F\colon\ F \in \mathcal{S}';,\ F(\varphi) < a\}$, where $\varphi \in \mathcal{S}$ and $a \in R^1$ are arbitrary. Given a probability space $(\mathcal{S}', \mathcal{A}_{\mathcal{S}'}, P)$, a generalized field $\bar{X} = \bar{X}(\varphi)$ can be defined on it by the formula $\bar{X}(\varphi)(F) = F(\varphi)$, $\varphi \in \mathcal{S}$, and $F \in \mathcal{S}'$. The following deep result is due to Minlos (see e.g. [16]).

Theorem (Minlos). *Let $(X(\varphi),\ \varphi \in \mathcal{S})$ be a generalized random field. There exists a probability measure P on the measurable space $(\mathcal{S}, \mathcal{A}_{\mathcal{S}'})$ such the generalized field $\bar{X} = (\bar{X}(\varphi),\ \varphi \in \mathcal{S})$ defined on the probability space $(\bar{S}, \mathcal{A}_{\mathcal{S}'}, P)$ by the formula $\bar{X}(\varphi)(F) = F(\varphi)$, $\varphi \in \mathcal{S}$, $F \in \mathcal{S}'$, satisfies the relation $X(\varphi) \stackrel{\Delta}{=} \bar{X}(\varphi)$ for all $\varphi \in \mathcal{S}$.*

The generalized field $\bar{X}$ has some nice properties. Namely property (a) in the definition of generalized fields holds for all $F \in \mathcal{S}'$. Moreover $\bar{X}$ satisfies the following strengthened version of property (b):

(b′) $\lim \bar{X}(\varphi_n) = \bar{X}(\varphi)$ in every point $F \in \mathcal{S}'$ if $\varphi_n \to \varphi$ in the topology of $\mathcal{S}$.

Because of this nice behaviour of the field $\bar{X}(\varphi)$ most authors define generalized fields as the versions $\bar{X}$ defined in Minlos' theorem. Since we have never needed the extra properties of the field $\bar{X}$ we have deliberately avoided the application of Minlos' theorem in the definition of generalized random fields. Minlos' theorem heavily depends on some topological properties of $\mathcal{S}$, namely that $\mathcal{S}$ is a so-called nuclear space. Minlos' theorem also holds if the space of test functions is substituted by $\mathcal{D}$ or $\mathcal{S}^r$ in the definition of generalized fields.

Let us finally remark that Lamperti [22] gave an interesting characterization of self-similar random fields. Let $X(t)$, $t \in R^1$, be a continuous time stationary random process, and define the random process $Y(t) = \frac{X(\log t)}{t^\alpha}$, $t > 0$, with some $\alpha > 0$. Then, as it is not difficult to see, the random processes $Y(t)$, $t > 0$, and $\frac{Y(ut)}{u^\alpha}$, $t > 0$, have the same finite dimensional distributions for all $u > 0$. This can be interpreted so that $Y(t)$ is a self-similar process with parameter $\alpha > 0$ on the half-line $t > 0$. Contrariwise, if the finite dimensional distributions of the processes $Y(t)$ and $\frac{Y(ut)}{u^\alpha}$, $t > 0$, agree for all $u > 0$, then the process $X(t) = \frac{X(e^t)}{e^{\alpha t}}$, $t \in R^1$, is stationary. These relations show some connection between stationary and self-similar processes. But they have a rather limited importance in the investigations of this work, because here we are really interested in such random fields which are simultaneously stationary and self-similar.

Chapter 2

Wick polynomials are widely used in the literature of statistical physics. A detailed discussion about Wick polynomials can be found in [12]. Theorems 2A and 2B are well-known, and they can be found in the standard literature. Theorem 2C can be found e.g. in Dynkin's book [14] (Lemma 1.5). Theorem 2.1 is due to Segal [32]. It is closely related to a result of Cameron and Martin [4]. The remarks at the end of the chapter about the content of formula (2.1) are related to [26].

Chapter 3

Random spectral measures were independently introduced by Cramer and Kolmogorov [5,21]. They could have been introduced by means of Stone's theorem about the spectral representation of one-parameter groups of unitary operators. Bochner's theorem can be found in any standard book on functional analysis, the proof of the Bochner–Schwartz theorem can be found in [16]. Let me remark that the same result holds true if the space of test functions $\mathcal{S}$ is substituted by $\mathcal{D}$.

There is an object, called the fractional Brownian motion, which is a popular topic of many investigations, and which can be studied by means of the method of this chapter. In particular, the results of Chap. 3 imply their existence. A fractional Brownian motion with Hurst parameter H, $0 < H < 1$, is a Gaussian process $X(t)$, $t \ge 0$, with continuous trajectories and zero expectation, i.e. $EX(t) = 0$ for all $t \ge 0$, and with covariance function $R_H(s,t) = EX(s)X(t) = \frac{1}{2}(s^{2H} + t^{2H} - |t-s|^{2H})$ for all $0 \le s,t < \infty$. Let us explain that the correlation of a fractional Brownian motion has a natural representation as the correlation function of the discretized version of an appropriately defined Gaussian stationary generalized self-similar field. In the subsequent argument the representation of (generalized) stationary Gaussian fields turned out to be very useful.

To find this representation observe that a fractional Brownian motion with Hurst parameter H has the self-similarity property $EX(as)X(at) = a^{2H}EX(s)X(t)$ for all $a > 0$, and simple calculation shows that it also has the following stationary increments property: $E[X(s+u) - X(u)][X(t+u) - X(u)] = EX(s)X(t)$ for all $0 \le s,t,u < \infty$. Hence we can construct a fractional Brownian motion $X(t)$ by defining first an appropriate stationary, Gaussian generalized self-similar field $\bar{X}(\varphi)$, $\varphi \in \mathcal{S}_1$ in the space of the real valued functions of the Schwartz space, and then by extending it to a larger parameter set (of functions), containing the indicator functions $\chi_{[0,t]}$ of the intervals $[0,t]$ for all $t \ge 0$. Finally we define the process $X(t)$ as $X(t) = \bar{X}(\chi_{[0,t]})$.

More explicitly, let us define for a parameter α the stationary generalized Gaussian field $\bar{X}(\varphi)$, $\varphi \in \mathcal{S}^1$, with zero expectation and spectral density $|u|^{-2\alpha}$, i.e. put $E\bar{X}(\varphi)\bar{X}(\psi) = \int \tilde{\varphi}(u)\bar{\tilde{\varphi}}(u)|u|^{-2\alpha}\,du$, and introduce its (discretized) extension to a function space containing the functions $\chi_{[0,t]}$ for all $t > 0$. Then we have

$$E\bar{X}(\chi_{[0,s]})\bar{X}(\chi_{[0,t]}) = \int \tilde{\chi}_{[0,s]}(u)\bar{\tilde{\chi}}_{[0,t]}(u)|u|^{-2\alpha}du = \int \frac{e^{isu}-1}{iu}\frac{e^{-itu}-1}{-iu}|u|^{-2\alpha}du,$$

provided that these integrals are convergent.

The above defined generalized fields exist if $2\alpha > -1$, and their discretized extension exists if $-1 < 2\alpha < 1$. The first condition is needed to guarantee that the singularity of the integrand in the formula expressing the covariance function is not too strong in the origin, and the second condition is needed to guarantee that the singularity of this integrand is not too strong at the infinity even in the discretized case.

Simple calculation shows that the covariance function of the above defined random field satisfies the identity $E\bar{X}(\varphi_a)\bar{X}(\psi_a) = a^{-(1+2\alpha)}E\bar{X}(\varphi)\bar{X}(\psi)$, with the functions $\varphi_a(x) = \varphi(ax)$, $\psi_a(x) = \psi(ax)$, and similarly, we have $EX(as)X(at) = a^{(1+2\alpha)}EX(s)X(t)$ for all $a > 0$. Besides, the Gaussian stochastic process $X(t)$, $t > 0$, has stationary increments, i.e. $E[X(s+u) - X(u)][X(t+u) - X(u)] = EX(s)X(t)$ for all $0 \le s,t,u < \infty$. This follows from its construction with the help of a stationary Gaussian random field.

The above calculations imply that with the choice $\alpha = H - 1/2$ we get the covariance function of a fractional Brownian motion with Hurst parameter H for all $0 < H < 1$, more precisely the correlation function of this process multiplied by an appropriate constant. Indeed, it follows from the stationary increments property of the process that $E(X(t) - X(s))^2 = EX(t-s)^2$, if $t \ge s$, and the self-similarity property of this process implies that $EX(s)X(t) = \frac{1}{2}[EX(s)^2 + EX(t)^2 - E(X(t) - X(s))^2] = \frac{1}{2}EX(1)^2[s^{2H} + t^{2H} - |t-s|^{2H}]$.

Actually the results of Chap. 3 also provide a representation of this process by means of an integral with respect to a random spectral measure. This representation has the form

$$X(t) = \int \frac{e^{itu} - 1}{iu} |u|^{-H+1/2} Z(du), \quad t > 0,$$

with the random spectral measure $Z(\cdot)$ corresponding to the Lebesgue measure on the real line. Here we omit the proof that such a stochastic process also has a version with continuous trajectories.

Chapter 4

The stochastic integral defined in this chapter is a version of that introduced by Itô in [19]. This modified integral first appeared in Totoki's lecture note [39] in a special form. Its definition is a little bit more difficult than the definition of the original stochastic integral introduced by Itô, but it has the advantage that the effect of the shift transformation can be better studied with its help. Most results of this chapter can be found in Dobrushin's paper [7]. The definition of Wiener–Itô integrals in the case when the spectral measure may have atoms is new. In the new version of this lecture note I worked out many arguments in a more detailed form than in the old text. In particular, in Lemma 4.1 I gave a much more detailed explanation of the statement that all kernel functions of Wiener–Itô integrals can be well approximated by simple functions.

Chapter 5

Proposition 5.1 was proved for the original Wiener–Itô integrals by Itô in [19]. Lemma 5.2 contains a well-known formula about Hermite polynomials. The main result of this chapter, Theorem 5.3, appeared in Dobrushin's work [7]. The proof given there is not complete. Several non-trivial details are omitted. I felt even necessary to present a more detailed proof in this note when I wrote down its new version. Theorem 5.3 is closely related to Feynman's diagram formula. The result of Corollary 5.5 was already known at the beginning of the twentieth century. It was proved with the help of some formal manipulations. This formal calculation was justified by Taqqu in [36] with the help of some deep inequalities. In the new version of this note I formulated a more general result than in the older one. Here I gave a formula about the expectation of products of Wick polynomials and not only of Hermite polynomials.

I could not find results similar to Corollaries 5.6 and 5.7 in the literature of probability theory. On the other hand, such results are well-known in statistical physics, and they play an important role in constructive field theory. A sharpened form of these results is Nelson's deep hypercontractive inequality [28], which I formulate below.

Let X_t, $t \in T$, and $Y_{t'}$, $t' \in T'$ be two sets of jointly Gaussian random variables on some probability spaces $(\Omega, \mathcal{A}, P)$ and $(\Omega, \mathcal{A}', P')$. Let $\mathcal{H}_1$ and $\mathcal{H}_1'$ be the Hilbert spaces generated by the finite linear combinations $\sum c_j X_{t_j}$ and $\sum c_j Y_{t_j'}$. Let us define the σ-algebras $\mathcal{B} = \sigma(X_t,\ t \in T)$ and $\mathcal{B}' = \sigma(Y_{t'},\ t' \in T')$ and the Banach spaces $L_p(X) = L_p(\Omega, \mathcal{B}, P)$, $L_p(Y) = L_p(\Omega', \mathcal{B}', P')$, $1 \le p \le \infty$. Let A be a linear transformation from $\mathcal{H}_1$ to $\mathcal{H}_1'$ with norm not exceeding 1. We define an operator $\Gamma(A)\colon L_p(X) \to L_{p'}(Y)$ for all $1 \le p, p' \le \infty$ in the following way. If η is a homogeneous polynomial of the variables X_t,

$$\eta = \sum C_{j_1,\dots,j_s}^{t_1,\dots,t_s} X_{t_1}^{j_1} \cdots X_{t_s}^{j_s}, \quad t_1, \dots, t_s \in T,$$

then

$$\Gamma(A)\colon\eta\colon = \sum C_{j_1,\dots,j_s}^{t_1,\dots,t_s} \colon (AX_{t_1})^{j_1} \cdots (AX_{t_s})^{j_s}\colon.$$

It can be proved that this definition is meaningful, i.e. $\Gamma(A)\colon\eta\colon$ does not depend on the representation of η, and $\Gamma(A)$ can be extended to a bounded operator from $L_1(X)$ to $L_1(Y)$ in a unique way. This means in particular that $\Gamma(A)\xi$ is defined for all $\xi \in L_p(X)$, $p \ge 1$. Nelson's hypercontractive inequality says the following. Let A be a contraction from $\mathcal{H}_1$ to $\mathcal{H}_1'$. Then $\Gamma(A)$ is a contraction from $L_q(X)$ to $L_p(Y)$ for $1 \le q \le p$ provided that

$$\|A\| \le \left(\frac{q-1}{p-1}\right)^{1/2}. \tag{9.1}$$

If (9.1) does not hold, then $\Gamma(A)$ is not a bounded operator from $L_q(X)$ to $L_p(Y)$.

A further generalization of this result can be found in [17].

The following discussion may help to understand the relation between Nelson's hypercontractive inequality and Corollary 5.6. Let us apply Nelson's inequality in the special case when $(X_t,\ t \in T) = (Y_{t'},\ t' \in T')$ is a stationary Gaussian field with spectral measure G, $q = 2$, $p = 2m$ with some positive integer m, $A = c \cdot \mathrm{Id}$, where Id denotes the identity operator, and $c = (2m-1)^{-1/2}$. Let $\mathcal{H}^c$ and $\mathcal{H}_n^c$ be the complexification of the real Hilbert spaces $\mathcal{H}$ and $\mathcal{H}_n$ defined in Chap. 2. Then

$L_2(X) = \mathcal{H}^c = \mathcal{H}_0^c + \mathcal{H}_1^c + \cdots$ by Theorem 2.1 and formula (2.1). The operator $\Gamma(c \cdot \mathrm{Id})$ equals $c^n \cdot \mathrm{Id}$ on the subspace $\mathcal{H}_n^2$. If $h_n \in \mathcal{H}_G^n$, then $I_G(h_n) \in \mathcal{H}_n$, hence the application of Nelson's inequality for the operator $A = c \cdot \mathrm{Id}$ shows that

$$\left(EI_G(h_n)^{2m}\right)^{1/2m} = c^{-n}\left(E(\Gamma(c\cdot\mathrm{Id})I_G(h_n))^{2m}\right)^{1/2m} \le c^{-n}\left(EI_G(h_n)^2\right)^{1/2}$$

i.e.

$$EI_G(h_n)^{2m} \le c^{-2nm}\left(EI_G(h_n)^2\right)^m = (2m-1)^{mn}\left(EI_G(h_n)^2\right)^m.$$

This inequality is very similar to the second inequality in Corollary 5.6, only the multiplying constants are different. Moreover, for large m these multiplying constants are near to each other. I remark that the following weakened form of Nelson's inequality could be deduced relatively easily from Corollary 5.6. Let $A\colon \mathcal{H}_1 \to \mathcal{H}_1'$ be a contraction $\|A\| = c < 1$. Then there exists a $\bar{p} = \bar{p}(c) > 2$ such that $\Gamma(A)$ is a bounded operator from $L_2(X)$ to $L_p(Y)$ for $p < \bar{p}$. This weakened form of Nelson's inequality is sufficient in many applications.

Chapter 6

Theorems 6.1, 6.2 and Corollary 6.4 were proved by Dobrushin in [7]. Taqqu proved similar results in [37], but he gave a different representation. Theorem 6.6 was proved by H.P. McKean in [27]. The proof of the lower bound uses some ideas from [15]. Remark 6.5 is from [25]. As Proposition 6.3 also indicates, some non-trivial problems about the convergence of certain integrals must be solved when constructing self-similar fields. Such convergence problems are common in statistical physics. To tackle such problems the so-called power counting method (see e.g. [23]) was worked out. This method could also be applied in this chapter. Part (b) of Proposition 6.3 implies that the self-similarity parameter α cannot be chosen in a larger domain in Corollary 6.4. One can ask about the behaviour of the random variables ξ_j and $\xi(\varphi)$ defined in Corollary 6.4 if the self-similarity parameter α tends to the critical value $\frac{\nu}{2}$. The variance of the random variables ξ_j and $\xi(\varphi)$ tends to infinity in this case, and the fields ξ_j, $j \in \mathbb{Z}_\nu$, and $\xi(\varphi)$, $\varphi \in \mathcal{S}$, tend, after an appropriate renormalization, to a field of independent normal random variables in the discrete, and to a white noise in the continuous case. The proof of these results with a more detailed discussion appeared in [10].

In a recent paper [20] Kesten and Spitzer have proved a limit theorem, where the limit field is a self-similar field which seems not to belong to the class of self-similar fields constructed in Chap. 6. (We cannot however, exclude the possibility that there exists some self-similar field in the class defined in Theorem 6.2 with the same distribution as this field, although it is given by a completely different form.) This self-similar field constructed by Kesten and Spitzer is the only rigorously constructed self-similar field known for us that does not belong to the fields constructed in Theorem 6.2. I describe this field, and then I make some comments.

Let $B_1(t)$ and $B_2(t)$, $-\infty < t < \infty$, be two independent Wiener processes. (We say that $B(t)$ is a Wiener process on the real line if $B(t)$, $t \geq 0$, and $B(-t)$, $t \geq 0$, are two independent Wiener processes.) Let $K(x, t_1, t_2)$, $x \in R^1$, $t_1 < t_2$, denote the local time of the process B_1 at the point x in the interval $[t_1, t_2]$. The one-dimensional field

$$Z_n = \int K(x, n, n+1)\, B_2(dx), \quad n = \dots, -1, 0, 1, \dots,$$

where the integral in the last formula is an Itô integral, is a stationary self-similar field with self-similarity parameter $\frac{3}{4}$.

To see the self-similarity property one has to observe that

$$K(\lambda^{1/2} x, \lambda t_1, \lambda t_2) \stackrel{\Delta}{=} \lambda^{1/2} K(x, t_1, t_2) \quad \text{for all } x \in R^1, \ \ t_1 < t_2, \text{ and } \lambda > 0$$

because of the relation $B_1(\lambda u) \stackrel{\Delta}{=} \lambda^{1/2} B_1(u)$. Hence

$$\sum_{j=0}^{n-1} Z_j \stackrel{\Delta}{=} n^{1/2} \int K(n^{-1/2} x, 0, 1) B_2(dx) \stackrel{\Delta}{=} n^{3/4} \int K(x, 0, 1)\, B_2(dx) = n^{3/4} Z_0.$$

The invariance of the multi-dimensional distributions of the field Z_n under the transformation (1.1) can be seen similarly.

To see the stationarity of the field Z_n we need the following two observations.

(a) $K(x, s, t) \stackrel{\Delta}{=} K(x + \eta(s), 0, t - s)$ with $\eta(s) = -B_1(-s)$. (The form of η is not important for us. What we need is that the pair (η, K) is independent of B_2.)
(b) If $\alpha(x)$, $-\infty < x < \infty$, is a process independent of B_2, then

$$\int \alpha(x+u) B_2(dx) \stackrel{\Delta}{=} \int \alpha(x) B_2(dx) \quad \text{for all } u \in R^1.$$

It is enough to show, because of Property (a) that

$$\int K(x + \eta(s), 0, t - s)\, B_2(dx) \stackrel{\Delta}{=} \int K(x, 0, t - s)\, B_2(dx).$$

This relation follows from property (b), because the conditional distributions of the left and right-hand sides agree under the condition $\eta(s) = u$, $u \in R^1$.

The generalized field version of the above field Z_n is the field

$$Z(\varphi) = -\int \left[K(x, 0, t) \frac{d\varphi}{dt}\, dt \right] B_2(dx), \quad \varphi \in \mathcal{S}.$$

To explain the analogy between the field Z_n and $Z(\varphi)$ we remark that the kernel of the integral defining Z_n can be written, at least formally, as

$$K(x, n, n+1) = \int \chi_{[n,n+1)}(u) \frac{d}{du} K(x, n, u)\, du,$$

although K is a non-differentiable function. Substituting the function $\chi_{[n,n+1)}$ by $\varphi \in \mathscr{S}$, and integrating by parts (or precisely, considering $\frac{d}{du} K$ as the derivative of a distribution) we get the above definition of $Z(\varphi)$.

Using the same idea as before, a more general class of self-similar fields can be constructed. The integrand $K(x, n, n+1)$ can be substituted by the local time of any self-similar field with stationary increments which is independent of B_2. Naturally, it must be clarified first that this local time really exists. One could enlarge this class also by integrating with respect to a self-similar field with stationary increments, independent of B_1. The integral with respect to a field independent of the field $K(x, s, t)$ can be defined without any difficulty.

There seems to be no natural way to represent the above random fields as random fields subordinated to a Gaussian random field. On the other hand, the local times $K(x, s, t)$ are measurable with respect to B_1, they have finite second moments, therefore they can be expressed by means of multiple Wiener–Itô integrals with respect to a white noise field. Then the process Z_n itself can also be represented via multiple Wiener–Itô integrals. It would be interesting to know whether the above defined self-similar fields, and probably a larger class of self-similar fields, can be constructed in a simple natural way via multiple Wiener–Itô integrals with the help of a randomization.

Chapter 7

The definition of Wiener–Itô integrals together with the proof of Theorem 7.1 and Proposition 7.3 were given by Itô in [19]. Theorem 7.2 is proved in Taqqu's paper [38]. He needed this result to show that the self-similar fields defined in [9] by means of Wiener–Itô integrals coincide with the self-similar fields defined in [38] by means of modified Wiener–Itô integrals.

Chapter 8

The results of this chapter, with the exception of Theorems 8.5 and 8.7 are proved in [9]. Theorem 8.5 is proved in [3] and Theorem 8.7 in [25]. The latter paper was strongly motivated by [30]. Lemma 8.3 is formulated in a more general form than Lemma 3 in [9]. The present formulation is more complicated, but it is more useful in some applications. Let me explain this in more detail. The difference between the original and the present formulation of this lemma is that here we allow that the integrand K_0 in the limiting stochastic integral is discontinuous on a small subset of $R^{k\nu}$, and the functions K_N may not converge on this set. This freedom can be exploited in some applications. Indeed, let us consider e.g. the self-similar fields constructed in Remark 6.5. In case $p < 0$ the integrand in the formula expressing these fields is not continuous on the hyperplane $x_1 + \cdots + x_n = 0$. Hence, if we want to prove limit theorems where these fields appear as the limit, and this happens e.g. in Theorem 8.7 then we can apply Lemma 8.3, but not its original version, Lemma 3 in [9].

The example for non-central limit theorems given by Rosenblatt in [29] and its generalization by Taqqu in [35] are special cases of Theorem 8.2. In these papers only the special case $H_2(x) = x^2 - 1$ is considered. Later Taqqu [38] proved a result similar to Theorem 8.2′, but he needed more restrictive conditions. The observation that Theorem 8.2′ can be deduced from Theorem 8.2 is from Taqqu [35].

The method of [29, 35] does not apply for the proof of Theorem 8.2 in the case of $H_k(x)$, $k \geq 3$. In these papers it is proved that the moments of the random variables Z_n^N converge to the corresponding moments of Z_n^*. (Actually a different but equivalent statement is established in these papers.) This convergence of the moments implies the convergence $Z_n^N \xrightarrow{\mathcal{D}} Z_n^*$ if and only if the distribution of Z_n^* is uniquely determined by its moments.

Theorem 6.6 implies that the $2n$-th moment of a k-fold Wiener–Itô integral behaves similarly to the $2kn$-the moment of a Gaussian random variable with zero expectation, it equals $e^{(kn \log n)/2 + O(n)}$. Hence some results about the so-called moment problem show that the distribution of a k-fold Wiener–Itô integral is determined by its moments only for $k = 1$ and $k = 2$. Therefore the method of moments does not work in the proof of Theorem 8.2 for $H_k(x)$, $k \geq 3$.

Throughout Chap. 8 I have assumed that the correlation function of the underlying Gaussian field to which our fields are subordinated satisfies formula (8.1). This assumption seems natural, since it implies that the spectral measure of the Gaussian field satisfies Lemma 8.1, and such a condition is needed when Z_{G_N} is substituted by Z_{G_0} in the limit. It can be asked whether in Theorem 8.2 formula (8.1) can be substituted by the weaker assumption that the spectral measure of the Gaussian field satisfies Lemma 8.1. This question was investigated in Sect. 4 of [9]. The investigation of the moments shows that the answer is negative. The reason for it is that the validity of Lemma 8.1, unlike that of Theorem 8.2, does not depend on whether the spectral measure G has large singularities outside the origin or not. The discussion in [9] also shows that the Gaussian case, that is the case when $H_k(x) = H_1(x) = x$ in Theorem 8.2, is considerably different from the non-Gaussian case. A forthcoming paper of M. Rosenblatt [31] gives a better insight into the above question.

The limiting fields appearing in Theorems 8.2 and 8.6 belong to a special subclass of the self-similar fields defined in Theorem 6.2. These results indicate that the self-similar fields defined in formula (6.7) have a much greater range of attraction if the homogeneous function f_n in (6.7) is the constant function. The reason for the particular behaviour of these fields is that the constant function is analytic, while a general homogeneous function typically has a singularity at the origin. A more detailed discussion about this problem can be found in [25].

References

1. Bleher, P.M., Missarov, M.D.: The equations of Wilson's renormalization group and analytic renormalization. Commun. Math. Phys. **74**(3), I. General results, 235–254, II. Solution of Wilson's equations, 255–272 (1980)
2. Bramson, M., Griffeath, D.: Renormalizing the three-dimensional voter model. Ann. Probab. **7**, 418–432 (1979)
3. Breuer, P., Major, P.: Central limit theorems for non-linear functionals of Gaussian fields. J. Multivar. Anal. **13**(3), 425–441 (1983)
4. Cameron, R.H., Martin, W.T.: The orthogonal development of nonlinear functionals in series of Fourier–Hermite functionals. Ann. Math. **48**, 385–392 (1947)
5. Cramer, H.: On the theory of stationary random processes. Ann. Math. **41**, 215–230 (1940)
6. Dawson, D., Ivanoff, G.: Branching diffusions and random measures. In: Advances in Probability. Dekker, New York (1979)
7. Dobrushin, R.L.: Gaussian and their subordinated generalized fields. Ann. Probab. **7**, 1–28 (1979)
8. Dobrushin, R.L.: Automodel generalized random fields and their renorm group. In: Multicomponent Random Systems. Dekker, New York (1980)
9. Dobrushin, R.L., Major, P.: Non-central limit theorems for non-linear functionals of Gaussian fields. Z. Wahrscheinlichkeitstheorie verw. Gebiete **50**, 27–52 (1979)
10. Dobrushin, R.L., Major, P.L: On the asymptotic behaviour of some self-similar fields. Sel. Mat. Sov. **1**(3), 293–302 (1981)
11. Dobrushin, R.L. Major, P., Takahashi, J.: Self-similar Gaussian fields. Finally it appeared as Major, P. (1982) On renormalizing Gaussian fields. Z. Wahrscheinlichkeitstheorie verw. Gebiete **59**, 515–533
12. Dobrushin, R.L., Minlos, R.A.: Polynomials of linear random functions. Uspekhi Mat. Nauk **32**, 67–122 (1977)
13. Dudley, R.M.: Distances of probability measures and random variables. Ann. Math. Stat. **39**, 1563–1572 (1968)
14. Dynkin, E.B.: Die Grundlagen der Theorie der Markoffschen Prozesse, Band 108. Springer, Berlin (1961)
15. Eidlin, V.L., Linnik, Yu.V.: A remark on analytic transformation of normal vectors. Theory Probab. Appl. **13**, 751–754 (1968) (in Russian)
16. Gelfand, I.M., Vilenkin, N.Ya.: Generalized Functions. IV. Some Applications of Harmonic Analysis. Academic (Harcourt, Brace Jovanovich Publishers), New York (1964)
17. Gross, L.: Logarithmic Soboliev inequalities. Am. J. Math. **97**, 1061–1083 (1975)
18. Holley, R.A., Stroock, D.: Invariance principles for some infinite particle systems. In: Stochastic Analysis, pp. 153–173. Academic Press, New York (1978)

P. Major, *Multiple Wiener-Itô Integrals*, Lecture Notes in Mathematics 849, DOI 10.1007/978-3-319-02642-8,

19. Itô, K.: Multiple Wiener integral. J. Math. Soc. Jpn. **3**, 157–164 (1951)
20. Kesten, H., Spitzer, F.: A limit theorem related to a class of self-similar processes. Z. Wahrscheinlichkeitstheorie verw. Gebiete **50**, 5–25 (1979)
21. Kolmogorov, A.N.: Wienersche Spirale und einige andere interessante Kurven im Hilbertschen Raum. C. R. (Doklady) Acad. Sci. U.R.S.S.(N.S.) **26**, 115–118 (1940)
22. Lamperti, J.: Semi-stable stochastic processes. Trans. Am. Math. Soc. **104**, 62–78 (1962)
23. Löwenstein, J.H., Zimmerman, W.: The power counting theorem for Feynman integrals with massless propagators. Commun. Math. Phys. **44**, 73–86 (1975)
24. Major, P.: Renormalizing the voter model. Space and space-time renormalization. Stud. Sci. Math. Hung. **15**, 321–341 (1980)
25. Major, P.: Limit theorems for non-linear functionals of Gaussian sequences. Z. Wahrscheinlichkeitstheorie verw. Gebiete **57**, 129–158 (1981)
26. McKean, H.P.: Geometry of differential space. Ann. Probab. **1**, 197–206 (1973)
27. McKean, H.P.: Wiener's theory of nonlinear noise (Proc. SIAM-AMS Sympos. Appl. Math., New York, 1972). In: Stochastic Differential Equations, SIAM-AMS Proc., vol. VI, pp. 191–209. Amer. Math. Soc. Providence, RI (1973)
28. Nelson, E.: The free Markov field. J. Funct. Anal. **12**, 211–227 (1973)
29. Rosenblatt, M.: Independence and dependence. In: Proceedings of Fourth Berkeley Symposium on Mathematical Statistics and Probability, pp. 431–443. University of California Press, Berkeley (1962)
30. Rosenblatt, M.: Some limit theorems for partial sums of quadratic forms in stationary Gaussian variables. Z. Wahrscheinlichkeitstheorie verw. Gebiete **49**, 125–132 (1979)
31. Rosenblatt, M.: Limit theorems for Fourier transform of functionals of Gaussian sequences. Z. Wahrscheinlichkeitstheorie verw. Gebiete **55**, 123–132 (1981)
32. Segal, J.E.: Tensor algebras over Hilbert spaces. Trans. Am. Math. Soc. **81**, 106–134 (1956)
33. Sinai, Ya.G.: Automodel probability distributions. Theory Probab. Appl. **21**, 273–320 (1976)
34. Sinai, Ya.G.: Mathematical problems of the theory of phase transitions. Akadémiai Kiadó, Budapest with Pergamon Press (1982)
35. Taqqu, M.S.: Weak convergence of fractional Brownian Motion to the Rosenblatt process. Z. Wahrscheinlichkeitstheorie verw. Gebiete **31**, 287–302 (1975)
36. Taqqu, M.S.: Law of the iterated logarithm for sums of non-linear functions of Gaussian variables. Z. Wahrscheinlichkeitstheorie verw. Gebiete **40**, 203–238 (1977)
37. Taqqu, M.S.: A representation for self-similar processes. Stoch. Process. Appl. **7**, 55–64 (1978)
38. Taqqu, M.S.: Convergence of iterated process of arbitrary Hermite rank. Z. Wahrscheinlichkeitstheorie verw. Gebiete **50**, 53–83 (1979)
39. Totoki, H.: Ergodic Theory. Lecture Note Series, vol. 14. Aarhus University, Aarhus (1969)

Index

P. Major, *Multiple Wiener-Itô Integrals*, Lecture Notes in Mathematics 849, DOI 10.1007/978-3-319-02642-8,

O

P

R

S

V

W

LECTURE NOTES IN MATHEMATICS Springer

Edited by J.-M. Morel, B. Teissier; P.K. Maini

Editorial Policy (for the publication of monographs)

1. Lecture Notes aim to report new developments in all areas of mathematics and their applications - quickly, informally and at a high level. Mathematical texts analysing new developments in modelling and numerical simulation are welcome.
   Monograph manuscripts should be reasonably self-contained and rounded off. Thus they may, and often will, present not only results of the author but also related work by other people. They may be based on specialised lecture courses. Furthermore, the manuscripts should provide sufficient motivation, examples and applications. This clearly distinguishes Lecture Notes from journal articles or technical reports which normally are very concise. Articles intended for a journal but too long to be accepted by most journals, usually do not have this "lecture notes" character. For similar reasons it is unusual for doctoral theses to be accepted for the Lecture Notes series, though habilitation theses may be appropriate.

2. Manuscripts should be submitted either online at www.editorialmanager.com/lnm to Springer's mathematics editorial in Heidelberg, or to one of the series editors. In general, manuscripts will be sent out to 2 external referees for evaluation. If a decision cannot yet be reached on the basis of the first 2 reports, further referees may be contacted: The author will be informed of this. A final decision to publish can be made only on the basis of the complete manuscript, however a refereeing process leading to a preliminary decision can be based on a pre-final or incomplete manuscript. The strict minimum amount of material that will be considered should include a detailed outline describing the planned contents of each chapter, a bibliography and several sample chapters.
   Authors should be aware that incomplete or insufficiently close to final manuscripts almost always result in longer refereeing times and nevertheless unclear referees' recommendations, making further refereeing of a final draft necessary.
   Authors should also be aware that parallel submission of their manuscript to another publisher while under consideration for LNM will in general lead to immediate rejection.

3. Manuscripts should in general be submitted in English. Final manuscripts should contain at least 100 pages of mathematical text and should always include
   - a table of contents;
   - an informative introduction, with adequate motivation and perhaps some historical remarks: it should be accessible to a reader not intimately familiar with the topic treated;
   - a subject index: as a rule this is genuinely helpful for the reader.

   For evaluation purposes, manuscripts may be submitted in print or electronic form (print form is still preferred by most referees), in the latter case preferably as pdf- or zipped ps-files. Lecture Notes volumes are, as a rule, printed digitally from the authors' files. To ensure best results, authors are asked to use the LaTeX2e style files available from Springer's web-server at:

   ftp://ftp.springer.de/pub/tex/latex/svmonot1/ (for monographs) and
   ftp://ftp.springer.de/pub/tex/latex/svmultt1/ (for summer schools/tutorials).

Additional technical instructions, if necessary, are available on request from lnm@springer.com.

4. Careful preparation of the manuscripts will help keep production time short besides ensuring satisfactory appearance of the finished book in print and online. After acceptance of the manuscript authors will be asked to prepare the final LaTeX source files and also the corresponding dvi-, pdf- or zipped ps-file. The LaTeX source files are essential for producing the full-text online version of the book (see http://www.springerlink.com/openurl.asp?genre=journal&issn=0075-8434 for the existing online volumes of LNM). The actual production of a Lecture Notes volume takes approximately 12 weeks.

5. Authors receive a total of 50 free copies of their volume, but no royalties. They are entitled to a discount of 33.3 % on the price of Springer books purchased for their personal use, if ordering directly from Springer.

6. Commitment to publish is made by letter of intent rather than by signing a formal contract. Springer-Verlag secures the copyright for each volume. Authors are free to reuse material contained in their LNM volumes in later publications: a brief written (or e-mail) request for formal permission is sufficient.

Addresses:

Professor J.-M. Morel, CMLA,
École Normale Supérieure de Cachan,
61 Avenue du Président Wilson, 94235 Cachan Cedex, France
E-mail: morel@cmla.ens-cachan.fr

Professor B. Teissier, Institut Mathématique de Jussieu,
UMR 7586 du CNRS, Équipe "Géométrie et Dynamique",
175 rue du Chevaleret
75013 Paris, France
E-mail: teissier@math.jussieu.fr

For the "Mathematical Biosciences Subseries" of LNM:

Professor P. K. Maini, Center for Mathematical Biology,
Mathematical Institute, 24-29 St Giles,
Oxford OX1 3LP, UK
E-mail: maini@maths.ox.ac.uk

Springer, Mathematics Editorial, Tiergartenstr. 17,
69121 Heidelberg, Germany,
Tel.: +49 (6221) 4876-8259

Fax: +49 (6221) 4876-8259
E-mail: lnm@springer.com